Framework for Technology Development and Delivery System in Agricultural Banking

Framework for Technology Development and Delivery System in Agricultural Bankingnt

Dr. R.P. Sharma

RANDOM PUBLICATIONS
NEW DELHI (INDIA)

Framework for Technology Development and Delivery System in Agricultural Banking

ISBN 978-93-5111-804-6

Published in 2016 in India by

RANDOM PUBLICATIONS

4376-A/4B, Gali Murari Lal, Ansari Road
New Delhi-110 002
Phone : +9111-43580356, 011-23289044, 011-43142548
e-mail: sales@randompublications.com,
info@randompublications.com, randomexports@gmail.com

Type Setting by : Friends Media, Delhi-110089
Digitally Printed at : Replika Press Pvt. Ltd.

Preface

The main sources of agriculture banking growth in India during 1956-87 were public agricultural research and extension; expansion of irrigated area and rural infrastructure and improvement in human capital were also important contributors.

The bank provides need-based agricultural loans of varying tenors to all credit worthy clients engaged in farming of staple as well as cash crops, horticulture, plantations, poultry, animal husbandry, dairying, seeds, warehousing, etc. The bank also finances the supplies of a wide range of agri inputs like seeds, fertilisers, pesticides, micro nutrients and micro irrigation tools. The bank has identified transportation, storage and processing of food and other agri commodities as a thrust area and is offering working capital and term loans of varying tenors to eligible processors, based on not only their financials but also on the strengths of underlying commodities.

Crop Loans are also called short term loans for "Seasonal Agricultural Operations." The Seasonal Agricultural Operations connote such activities as are undertaken in the process of raising various crops and are seasonally recurring in nature. The activities include, among others, ploughing and preparing land for sowing, weeding, transplantation where necessary, acquiring and applying inputs such as seeds, fertilizers, insecticides etc. and labour for all operations in the field for raising & harvesting the crops. Thus, the credit required to meet the current expenditure for raising the crops on land till the crops are harvested is construed as production or short term credit for seasonal agricultural operations.

This book makes an attempt to present the available information on the modern concepts of agricultural research and advanced principles in crop production easily understandable manner.all covering the latest syllabus prescribed by most of the Indian universities.

– Author

Contents

1

Agricultural Business and Farming Management

FARMING AS A BUSINESS

Farming has not usually been considered a business. The diversity of the duties of the farmer, the area over which the operations of the farm extend and the complexity of the records required, combine in making difficult the organization of the details of farming into business form. The successful financial operation of a farm presents quite as complex problems and calls for at least as much business ability and judgment as is required in operating a store with the same investment.

Farming, therefore, should be considered as a business, and the man who can produce his crops and products at the lowest cost and sell them at the highest price, investing the proceeds to the best advantage, should be considered the best farm manager. The man who knows the details of the cost of production and operation, and whose records show the profitable and unprofitable lines of production, thus enabling him to eliminate those that do not yield a profit, may be counted as the best business man.

A farmer should know the elements of soil fertility. He must understand the principles of the movement of soil water, and the action of soil bacteria. He should understand the nature of plant growth and be familiear with varieties and species of plants and with the effect of one crop on the crop following. He must also be familiar with animals and their habits and know how to feed and care for them. In addition, he must know how to buy and sell to advantage, make contracts, and plan his buildings and his farm so as to necessitate a light expenditure for labour, also that he may distribute his labour to advantage over the various farm enterprises. And he should know how to keep accounts.

The farmer in organizing his business could well follow the example of the merchant. The merchant first takes an inventory of his stock. He studies the demand for his goods, both present and prospective. He notes the supply, the cost, and the demand for each article. He calculates the labour required to operate his business and such other items of expense are considered as may be legitimately charged against the business.

He regulates his purchases and his prices according to the cost of securing his goods and putting them on the market. In conducting a large store business, it is customary to organize it into departments, putting some competent person in charge of each department and having the labour and accounting charges so systematized and recorded as to show the profit or loss from each department and from the business as a whole.

The farmer should likewise take an inventory of his capital, stock, and equipment. He should consider the type of farming to which the soil and climate are adapted. He should consider the fertility of the soil and the demand that will be made upon it by the crops grown. He should consider, in connection with the soil fertility, the sources from which it may be renewed and at what cost. He must study the markets, the transportation, and the demand for such crops as he grows; also the cost of producing each of the crops and the probable net profit that will be returned. His labour likewise should be charged against the various crops or enterprises and distributed to the best advantage.

In studying the problems of farm organizations, interest on investment, taxes, insurance, and other expense must be included as they affect the financial result. As in a large store business, it is frequently necessary to organize the large farm into departments, keeping accounts with the dairy, with the swine, the grain crops, the garden, and other similar enterprises.

Where the business is large enough, it is well to put an expert in charge of each large branch or group of enterprises, thus enabling one to use cheaper labour for performing the work or making the labour more effective by closer supervision. Where the farming is conducted as an organized business, and accounts are kept with the various lines of work, it is possible at the end of the year, to make a business statement which will show which lines have been profitable. The manager then can change his methods or drop out those lines that prove to be unprofitable and the business as a whole may be put on a better basis.

INVESTMENT

The investment of money in land, buildings, and equipment demands careful consideration. It is possible to pay so much for a farm that it will be impossible to produce sufficient revenue to meet the expense of operation and to pay a normal rate of interest on the money invested. This is particularly true where low-priced products are produced. A farm may be highly productive but so located that it will be impossible to market the produce on a profit bearing basis. One should study closely the market facilities of the neighbourhood and raise supplies which can be successfully marketed locally, or which can be transported to a market that pays well for such produce. Unless the produce of the farm is well related to the market, the farm is likely to be operated at a loss.

PROPORTION IN REAL ESTATE

It is a mistake for one to invest all of his capital in the real estate itself. Sufficient capital should be reserved for operating the farm. The hunger for land has induced many farmers to buy more land that they can equip and operate well. Such farmers are said to be land poor. They would secure greater profit from a medium sized farm, well tilled and managed, than from a large one which is insufficiently equipped and poorly operated. Rarely should more than 50 or 60 per cent of the capital be tied up in the land. The size of the farm and the amount of equipment are closely related to the possible profits.

A farm of forty or eighty acres devoted to diversified crops and live stock, cannot be so economically equipped per acre as a larger farm. The investment per acre in machinery will be higher as the cost will be spread over fewer acres than the machinery has capacity to handle. Investment in other equipment will also be correspondingly high. Often the labour on such a farm is not fully employed and loss results from inactivity of labour and equipment. A medium to large sized farm, when well organized, fully equipped, and with sufficient capital reserved to operate it well, will pay a much better labour income than a small farm.

ECONOMY OF LANDLORD FARMING IN THE AREA DESCRIBED

We have cited the evidence of agronomists and farmers to the effect that dairy farming on the landlord estates leads to the rationalisation of agriculture. Let us add here that the analysis of the Zemstvo statistics on this question made by Mr. Raspopin fully confirms this conclusion. We refer the reader to Mr. Raspopin's article for detailed data and give here only his main conclusion. "The interdependence of the condition of stock raising and dairy farming, on the one hand, and the number of dilapidated estates and the intensity of farming, on the other, is beyond question. The uyezds (of Moscow Gubernia) where dairy cattle raising, dairy farming, is most developed show the smallest percentage of dilapidated farms and the highest percentage of estates with highly developed field cultivation. Throughout Moscow Gubernia ploughland is being reduced and turned into meadow and pastureland, while grain rotations are yielding place to multi-field herbage rotations. Fodder grasses and dairy cattle, and not grain, are now predominant... not only on the farming estates in Moscow Gubernia but throughout the Moscow industrial district".

SCALE OF BUTTER PRODUCTION AND CHEESE MAKING

The scale of butter production and cheese making is particularly important precisely because it testifies to a complete revolution in agriculture, which becomes entrepreneur farming and breaks with routine. Capitalism subordinates to itself one of the products of agriculture, and all other aspects of farming are fitted in with this principal product. The keeping of dairy cattle calls forth the cultivation of grasses, the change-over from the three-field system to multi-field systems, etc. The waste products of cheese making go to fatten cattle for the market. Not only milk processing, but the whole of agriculture becomes a commercial enterprise. The influence of cheese production and butter making is not confined to the farms on which they are carried on, since milk is often bought up from the surrounding peasants and landlords. By buying up the milk, capital subordinates to itself the small agriculturists too, particularly with the organisation of the so-called "amalgamated dairies," the spread of which was noted in the 70s.

These are establishments organised in big towns, or in their vicinity, which process very large quantities of milk brought in by rail. As soon as the milk arrives the cream is skimmed and sold fresh, while the skimmed milk is sold at a low price to poorer purchasers. To ensure that they get produce of a certain quality, these establishments sometimes conclude contracts with the suppliers, obliging them to adhere to certain rules in feeding their cows. One can easily see how great is the significance of large establishments of this kind: on the one hand they capture the public market (the sale of skimmed milk to the poorer town-dwellers), and on the other hand they enormously expand the market for the rural entrepreneurs. The latter are given a tremendous impetus to expand and improve commercial farming. Large-scale industry brings them into line, as it were, by demanding produce of a definite quality and forcing out of the market (or placing at the mercy of the usurers) the small producer who falls below the "normal" standard. There should also operate in the same direction the grading of milk as to quality (fat content, for example), on which technicians are so busily engaged, inventing all sorts of lacto-densimeters, etc., and of which the experts are so heartily in favour.

Commercial Grain Farming

In this respect the role of the amalgamated dairies in the development of capitalism is quite analogous to that of elevators in commercial grain farming. By sorting grain as to quality the elevators turn it into a product that is not individual but generic (*res fungibilis*, as the lawyers say), *i.e.*, for the first time they adapt it fully to exchange. Thus, the elevators give a powerful impetus to commodity-grain production and spur on its technical development by also introducing grading for quality. Such a system strikes a double blow at the small producer. Firstly, it sets up as a standard, legalises, the higher-quality grain of the big crop sowers and thereby greatly depreciates the inferior grain of the peasant poor. Secondly, by organising the grading and storing of grain on the lines of large-scale capitalist industry, it reduces the big sowers' expenses on this item and facilitates and simplifies the sale of grain for them, thereby placing the small producer, with his patriarchal and primitive methods of selling from the cart in the market, totally at the mercy of the kulaks and the usurers. Hence, the rapid development of elevator construction in recent years means as big a victory for capital and degradation of the small commodity-producer in the grain business as does the appearance and development of capitalist "amalgamated dairies."

From the foregoing material it is clear that the development of commercial stock farming *creates* a home market, firstly, for means of production—milk-processing equipment, premises, cattle sheds, improved agricultural implements required for the change-over from the routine three-field system to multi-field crop rotations, etc.; and secondly, for labour-power. Stock farming placed on an industrial footing requires a far larger number of workers than the old stock farming "for manure." The dairy farming area—the industrial and north-western gubernias—does really attract masses of agricultural labourers. Very many people go to seek agricultural work in the Moscow, St. Petersburg, Yaroslavl and Vladimir gubernias; fewer, but nevertheless a considerable number, go to the Novgorod, Nizhni Novgorod and other non-black-earth gubernias. According to correspondents of the Department of Agriculture in the Moscow and other gubernias private-landowner farming is actually conducted in the main by labourers from other areas. This paradox— the migration of agricultural workers from the agricultural gubernias (they come mostly from the central black-earth gubernias and partly from the northern) to the industrial gubernias to do agricultural jobs in place of industrial workers who abandon the area en masse—is an extremely characteristic phenomenon.

It proves more convincingly than do any calculations or arguments that the standard of living and the conditions of the working people in the central black-earth gubernias, the least capitalist ones, are incomparably lower and worse than in the industrial gubernias, the most capitalist ones; it proves that in Russia, too, the following has become a universal fact, namely, the phenomenon characteristic of all capitalist countries, that the conditions of the workers in industry are better than those of the workers in agriculture (because in agriculture oppression by capitalism is supplemented by the oppression of pre-capitalist forms of exploitation). That explains the flight from agriculture to industry, whereas not only is there no flow from the industrial gubernias towards agriculture (for example, there is no migration from these gubernias at all), but there is even a tendency to look down upon the "raw" rural workers, who are called "cowherds" (Yaroslavl Gubernia), "cossacks" (Vladimir Gubernia) and "land labourers" (Moscow Gubernia).

It is important also to note that cattle herding requires a larger number of workers in winter than in summer. For that reason, and also because of the development of agricultural processing trades, the demand for labour in the area described not only grows, but is *more evenly distributed over the whole year* and over a period of years. The most reliable material for judging this interesting fact is the data on wages, if taken for a number of years. We give these data, confining ourselves to the groups of Great-Russian and Little-Russian gubernias. We omit the western gubernias, owing to their specific social conditions and artificial congestion of population (the Jewish pale of settlement), and quote the Baltic gubernias only to illustrate the relations that arise where capitalism is most highly developed.

Let us examine this table, in which the three principal columns are printed in italics. The first column shows the proportion of summer to yearly pay. The *lower* this proportion is, and the nearer the summer pay approximates to half the yearly pay, the more evenly is the demand for labour spread over the entire year and *the less the winter unemployment*. The least favourably placed in this respect are the central black-earth gubernias—the area where labour-service prevails and where capitalism is poorly developed. In the industrial gubernias, in the dairy-farming area the demand for labour is higher and winter unemployment is less. Over a period of years, too, the pay is most stable here, as may be seen from the second column, which shows the difference between the lowest and the highest pay in the harvest season. Lastly, the difference between the pay in the sowing season and the pay in the harvest season is also least in the non black-earth belt, *i.e.*, the demand for workers is more evenly distributed over the spring and summer. In all respects mentioned the Baltic gubernias stand even higher than the non-black-earth gubernias, while the steppe gubernias, with their immigrant workers and with harvest fluctuations of the greatest intensity, are marked by the greatest instability of wages. Thus, the data on wages testify that agricultural capitalism in the area described not only creates a demand for wage-labour, but also distributes this demand more evenly over the whole year.

Lastly, reference must be made to one more type of dependence of the small agriculturist in the area described upon the big farmer. This is the replenishment of landlords' herds by the purchase of cattle from peasants. The landlords find it more profitable to buy cattle from peasants driven by need to sell "at a loss" than to breed cattle themselves—just as our buyers-up in so-called handicraft industry often prefer to buy finished articles from the handicraftsmen at a ruinously cheap price rather than manufacture them in their own workshops.

This fact, which testifies to the extreme degradation of the small producer, and to his being able to keep going in modern society only by endlessly reducing his requirements, is turned by Mr. V. V. into an argument in favour of small "people's" production!... "We are entitled to draw the conclusion that our big farmers... do not display a sufficient degree of independence.... The peasant, however... reveals greater ability to effect real farming improvements". This lack of independence is expressed in the fact that "our dairy farmers... buy up the peasants' (cows) at a price rarely amounting to half the cost of raising them—usually at not more than a third, and often even a quarter of this cost". The merchant's capital of the stock farmers has made the small peasants completely dependent, it has turned them into its cowherds, who breed cattle for a mere song, and has turned their wives into its milkmaids. One would think that the conclusion to be drawn from this is that there is no sense in retarding the transformation of merchant's capital into industrial capital, no sense in supporting small production, which leads to forcing down the producer's standard of living below that of the farm labourer. But Mr. V. V. thinks otherwise. He is delighted with the "zeal" of the peasant in tending his cattle; he is delighted with the "good results from livestock farming" obtained by the peasant woman who "spends all her life with her cow and sheep". What a blessing, to be sure! To "spend all her life with her cow" (the milk of which goes to the improved cream separator), and as a reward for this life, to receive "one-fourth of the cost" of tending this cow! Now really, how after that can one fail to declare in favour of "small people's production"!

COSMETIC THE PRODUCTIVITY OF SMALL FARMING SYSTEMS

Despite the evidence of the resiliency and productivity advantages of small-scale and traditional farming systems, many scientists and growth specialists and organizations argue that the performance of subsistence agriculture is unsatisfactory, and that agrochemical and transgenic intensification of production is essential for the transition from subsistence to commercial production. Although such intensification approaches have met with much failure, research indicates that traditional crop and animal combinations can often be adapted to increase productivity. This is the case when ecological principles are used in the redesign of small farms, enhancing the habitat so that it promotes healthy plant growth, stresses pests, and encourages beneficial organisms while using labour and local resources more efficiently.

Several reviews have amply documented that small farmers can produce much of the needed food for rural and neighbouring urban communities in the midst of climate change and burgeoning energy costs. The evidence is conclusive: new agroecological approaches and technologies spearheaded by farmers, NGOs, and some local governments around the world are already making a sufficient contribution to food security at the household, national, and regional levels. A variety of agroecological and participatory approaches in many countries show very positive outcomes even under adverse environmental conditions. Potentials include: raising cereal yields from 50 to 200 percent, increasing stability of production through diversification, improving diets and income, and contributing to national food security (and even to exports) and conservation of the natural resource base and biodiversity. This evidence has been reinforced by a recent report of the United Nations Conference on Trade and Growth stating that organic agriculture could boost African food security.

Based on an analysis of 114 cases in Africa, the report revealed that a conversion of farms to organic or near-organic production methods increased agricultural productivity by 116 percent. Moreover, a shift towards organic production systems has enduring impact, as it builds up levels of natural, human, social, financial, and physical capital in farming communities. The International Assessment of Agricultural Knowledge, Science and Technology (AKST) commissioned by World Bank and the Food and Agriculture Organization (FAO) of the United Nations recommended that an increase and strengthening of AKST towards agroecological sciences will contribute to addressing environmental issues while maintaining and increasing productivity. The assessment also stresses that traditional and local knowledge systems enhance agricultural soil quality and biodiversity as well as nutrient, pest, and water management, and the capacity to respond to environmental stresses such as climate.

Whether the potential and spread of agroecological innovations is realized depends on several factors and major changes in policies, institutions, and research and growth approaches. Proposed agroecological strategies need to target the poor deliberately, and not only aim at increasing production and conserving natural resources. But they must also create employment and provide access to local inputs and local markets. Any serious attempt at developing sustainable agricultural technologies must bring to bear local knowledge and skills on the research process. Particular emphasis must be given to involving farmers directly in the formulation of the research agenda and on their active participation in the process of technological innovation and dissemination through Campesino a Campesino models that focus on sharing expcriences, intensification local research, and problem-solving capacities. The agroecological process requires participation and enhancement of the farmer's ecological literacy about their farms and resources, laying the foundation for empowerment and continuous innovation by rural communities. Equitable market opportunities must also be developed, emphasizing local commercialization and distribution schemes, fair prices, and other mechanisms that link farmers more directly and with greater solidarity to the rest of the population. The ultimate challenge is to increase investment and research in agroecology and scale up projects that have already proven successful to thousands farmers. This will generate a meaningful impact on the income, food security, and environmental well-being of all the population, especially small farmers who have been adversely impacted by conventional modern agricultural policy, technology, and the penetration of multinational agribusiness deep into the third world.

RURAL SOCIAL MOVEMENTS, AGROECOLOGY, AND FOOD SOVEREIGNTY

The growth of sustainable agriculture will require significant structural changes, in addition to technological innovation, farmer-to-farmer networks, and farmer-to-consumer solidarity. The required change is impossible without social movements that create political will among decision-makers to dismantle and transform the institutions and regulations that presently hold back sustainable agricultural growth. A more radical transformation of agriculture is needed, one guided by the notion that ecological change in agriculture cannot be promoted without comparable changes in the social, political, cultural, and economic arenas that help determine agriculture. The organized peasant and indigenous-based agrarian movements—such as the international peasant movement La Vía Campesina and Brazil's Landless Peasant Movement (MST)—have long argued that farmers need land to produce food for their own communities and for their country. For this reason they have advocated for genuine agrarian reforms to access and control land, water, and biodiversity that are of central importance for communities in order to meet growing food demands.

Vía Campesina believes that in order to protect livelihoods, jobs, people's food security, and health as well as the environment, food production has to remain in the hands of small-scale sustainable farmers and cannot be left under the control of large agribusiness companies or supermarket chains. Only by changing the export-led, free-trade based, industrial agriculture model of large farms can the downward spiral of poverty, low wages, rural-urban migration, hunger, and environmental degradation be halted. Social rural movements embrace the concept of food sovereignty as an alternative to the neoliberal approach that puts its faith in an inequitable international trade to solve the world's food problem. Instead, it focuses on local autonomy, local markets, local production-consumption cycles, energy and technological sovereignty, and farmer-to-farmer networks.

"Greening" the Green Revolution will not be sufficient to reduce hunger and poverty and conserve biodiversity. If the root causes of hunger, poverty, and inequity are not confronted head-on, tensions between socially equitable growth and ecologically sound conservation are bound to accentuate. Organic farming systems that do not challenge the monoculture nature of plantations and rely on external inputs as well as foreign and expensive certification seals, or fair-trade systems destined only for agro-export, offer very little to small farmers that become dependent on external inputs and foreign and volatile markets. By keeping farmers dependent on an input substitution approach to organic agriculture, fine-tuning of input use does little to move farmers toward the productive redesign of agricultural ecosystems that would move them away from dependence on external inputs. Niche markets for the rich in the North exhibit the same problems of any agro-export scheme that does not prioritize food sovereignty, perpetuating dependence and hunger. Rural social movements understand that dismantling the industrial agrifood complex and restoring local food systems must be accompanied by the construction of agroecological alternatives that suit the needs of small-scale producers and the low-income non-farming population, and that oppose corporate control over production and consumption. Given the urgency of the problems affecting agriculture, coalitions that can rapidly foster sustainable agriculture among farmers, civil society organizations (including consumers), as well as relevant and committed research organizations are needed. Moving toward a more socially just, economically viable, and environmentally sound agriculture will be the result of the coordinated action of emerging social movements in the rural sector in alliance with civil society organizations that are committed to supporting the goals of these farmers movements. As a result of constant political pressure from organized farmers and others, politicians will, it is hoped, become more responsive to developing policies that will enhance food sovereignty, preserve the natural resource base, and ensure social equity and economic agricultural viability.

PROPORTION OF INVESTMENT IN MACHINERY

Investments in machinery are worthy of quite as much consideration as investments in land. Machinery is looked upon as one of the means of reducing the cost of production. The wise use of machinery saves time and labour and enables the farmer to handle large acreages. In this light, the use of ample machinery is wise. The fact remains, however, that investments in machinery are often poorly made and that many farmers are embarrassed by debts for machinery. Frequently, farmers purchase machinery because it is fashionable or because a neighbour has it, rather than because carefully made calculations show that a certain machine can be used profitably. A safe rule is to buy no machine until carefully made calculations show that the cost of production of a certain crop or product will be reduced sufficiently by the purchase to cover the cost of the machine. That machinery investments can be studied from a business standpoint is quite plain. The following example, showing the relative cost of cutting corn with a machine versus cutting by hand, will illustrate:

The original cost of a corn binder is $125. The annual depreciation, as shown by statistical records covering ten years' work with farmers in Minnesota, is $12.50. The interest on the investment at the average value of the binder throughout its life will be $4.12. Repairs, shelter, and insurance will cost $2 annually. The total annual cost for the use of the binder, therefore, will be $18.62. If only twenty acres of corn are grown each year, the annual cost an acre for the corn binder will be $.93. The cost of cutting corn would be as follows: It will be noted that a twenty-acre corn field can be more economically harvested by hand where labour is available.

On a ten-acre field, the difference in expense would be still greater in favour of the hand harvesting, as the cost of machinery per acre would be doubled. The scarcity of labour, however, and the necessity of harvesting corn quickly to save it from frost, would often warrant the expenditure for the machine, even though it does slightly raise the cost of harvesting per acre. The greater the acreage, the more useful the machine becomes in harvesting, and the less the expense per acre for machinery use. Calculations similar to this should be made before purchasing a machine for any purpose. If it can be shown that the cost of performing the labour may be reduced by the machine, and that the labour can be performed in a more satisfactory and efficient manner with the use of it, then the purchase may be warranted. In many cases, however, calculation will show that the purchase is not warranted and that it would be better to hire labour and rent the machine, or to buy in partnership with some one else.

COST OF MOTIVE POWER

Another factor that should receive consideration is the cost of motive power in use on the farm. Horses usually furnish the farm motive power, though tractors are used to advantage in some cases. Auto trucks and automobiles can often be used to advantage in marketing dairy, fruit and garden products and the cost of using them, when it can be ascertained, should be compared with the cost of horse power. It costs from $40 to $100 per year to keep a work horse, depending on the locality and on the price of feed stuffs; also on the work that the horse does. The average cost a year of keeping a horse in Minnesota for the years 1904 to 1907, varied from $75.07 at Halstad, to $90.40 at Northfield.

Often a large number of horses are kept because the farm has been devoted to grain raising and the horses are needed at seeding and harvest times. They run in the pasture during the summer and are idle during the winter months. They must be fed and cared for during this time. The money invested in them would be drawing interest if invested somewhere else. Farmers should reduce the horses kept to the number actually required to do the work, unless the surplus are colts growing in value as one of the market products of the farm. The work of a farm can often be lessened by adopting a good crop rotation and using such crops as do not demand large amounts of horse labour at the same time. In this way a farm of 160 to 240 acres can often be worked with four to six horses, whereas eight to ten are frequently kept. The support of two or three extra horses per year would amount to $200 to $250 and is an item well worth saving.

INVESTMENT IN BUILDINGS

Investment in buildings, fences, and other items of equipment should be considered in the same business-like way. A barn costing $4,000 and providing shelter for forty head of cattle would carry with it an annual cost of $440. This annual cost is made up from the interest on the investment, insurance, depreciation, paint, and repairs. It will be about 11 per cent on the total investment. If the same forty cattle could be housed in a barn costing $2000 the total annual cost would be only $220, charging the same interest and expense rates as in the first instance, and assuming that the rate of depreciation would be the same.

While the $4000 barn would undoubtedly be a better barn, it would not add to the production of the cows housed, unless it was much more comfortable. It would not add to the net profit from the investment unless the labour of doing the chores and caring for the cows would be considerably reduced by greater convenience. The cost of horse barns, swine barns, and other buildings can be similarly calculated. One should not erect a building unless it is going to add to the efficiency of the live stock, shelter hay or machinery, or lessen the labour of doing the chores. It is wiser to invest money in drainage or in better tillage of the soil, than to invest it in buildings that shelter unproductive stock, or that add nothing to the earning power of the farm.

COST OF LABOUR

The employment, organization and direction of labour demands considerable study. The value of a good farm manager lies quite as much in his abilitiy so to select and direct labour as to yield a profit, as it does in his ability to drive a good bargain or sell his crops well. The only reason for employing labour is to increase the product and consequent profit. If a farmer can, by employing a man eight months in the year at $40 per month, increase the product of his farm by $500, he will be warranted in employing the labour. If, however, the $320 invested in labour should yield an increase of only $200 in the products of the farm, employment would be at a loss.

FACTORS OF PRODUCTION

Three primary factors are necessary in agricultural production. These are capital, land, and labour. The adjustment of these three factors is an important part of the business of the farm owner or manager, and determines largely the profits that may be made from the individual farm.

Capital, as commonly understood, includes the money value represented in the investment of the farm property, no matter what the form may be. Implements, live stock, teams, buildings, and othe rarticles of equipment, ar each a part of the capital of the farm. Cash for operating is also included.

*Good buildings and fences and a well kept farm often help in attracting customers for stock, seed grain or other products. As an advertisement they may increase the earning power of the farm indirectly. The satisfaction of owning good buildings and their influence in keeping the young folks on the farm or in enabling one to keep hired help should also be considered. The reputation of a farm in this respect may become a business asset.

Land represents the larger part of capital on most farms, and demands special consideration because the amount of land available for agricultural purposes is limited. Farmers have for this reason, regarded it wise to secure large quantities in localities where it was cheap, anticipating a rise in value. Location and demand for land in particular sections has led to much speculation, and land values fluctuate frequently.

The proportionate investment in each of the three forms,—circulating capital, land, and labour—bears a vital relation to the profits possible from the farm, and must be given the most careful consideration by the person who is buying and equipping a farm.

CAPITAL CLASSIFIED

There are two forms of capital in common use. They are known as fixed or invested capital, and circulating or working capital. The fixed capital properly includes all forms of permanent equipment, such as investment in land, buildings, implements, teams, and other articles that are used continuously. In land it includes the natural value and the value of the improvements that have been made upon it. Picking stones from a rough section of land adds to its value and increases the capital invested. Clearing trees from the land has the same effect. Wells, drainage, roads, fences, and other forms of improvement which are permanent and which become part of the land, also add to the natural value and become a part of the fixed capital. Buildings also are looked upon as part of the fixed capital. Strictly speaking, only those buildings which add to the producing power of the farm should be included in the capital invested in the farm. The dwelling house, while commonly added to the investment in the farm, is really intended for the personal use of the farmer and his family. Except in so far as it shelters the help employed on the farm, it can add but little to the returns from it. So far as making a statement of the business of the farm is concerned, it would be better were the farm-house inventoried separately from the other buildings and regarded as a personal expense to the farmer, just as the house of the banker in the city is separated from the business of the bank. All other buildings including silos, corn cribs, granaries, and buildings for sheltering the stock and necessary in conducting the farm business, should be included in the inventoried capital of the farm.

Equipment in the way of teams for work purposes; implements; live stock, such as cows, brood sows, sheep and poultry, that are kept for live stock products, are all a part of the permanent equipment, since they are permanently employed and if sold are replaced by other animals for the same purpose.

The circulating or working capital, includes such items of equipment as are frequently changing. Seed grain, household and farm supplies that are immediately used or marketed, live stock, such as fattening steers, and money for hired labour, are examples of circulating capital. The classificiation intends tha the term "circulating capital" shall include only those items that are used once and disappear. If sold for cash, the cash may be invested in other forms of working capital which in turn disappear. Needless to say, the amount of working or circulating capital varies greatly in accordance with the type of business done, with the market, and with the tastes of the farmer. No rule can be given for the exact adjustment of capital for these reasons.

PRODUCTION LIMITED BY DEFICIENT FACTOR

It is a common experience that the production on a farm is limited by the minimum amount of the one deficient factor. Difficulty is experienced in securing profitable production on a limited land area. In such a case, labour will not be fully employed or the equipment cannot be used effectively. On the other hand, a large land area and large equipment cannot be used to advantage without a plentiful supply of labour. Again, neither land nor labour can be used to the best advantage if the equipment is inadequate. It therefore stands to reason that these three factors must be carefully considered and proportioned in accordance with the needs of the business.

In purchasing and organizing a farm with limited capital, it is believed best to make the investment in about the following proportions:

- 45% for land investment
- 20% in buildings, provided they are to shelter productive live stock or market products;
- 22% in work animals and live stock;
- 8% in implements and tools;
- 5% reserved as working capital.

In buying a farm and equipping it with new machinery, the land investment will run lower than above indicated and the implements and tools investment will run higher. The machinery, however, will depreciate in value while the land is more likely to increase in value.

As the farm becomes older and the land is improved, the proportion of investments will gradually change. On old farms near city markets, the machinery investment may become comparatively insignificant.

EXERCISES FOR PUPILS

Farm Inventory

Have the pupils take inventories of their fathers' farms, providing a form similar to the following. The inventory may be taken on the regular weekly holiday. Prices should be checked over by the proprietor of the farm and brought to class for discussion and completion. The teacher should, if possible, go with the class to some farm and take an inventory, with the help of the proprietor, before starting the class members on their individual work. Allowing for any withdrawals of or additions to capital, the difference between the totals of two successive inventories will be the gain.

Proportion of Investment

Have the pupils find the total investment and the per cent of the capital invested in each of the forms of equipment used in classifying the inventory.

Cost of Shelter

Have the pupils learn the cost of the barns on their fathers' farms. Ask them to calculate the interest at the prevailing rate. Determine the amount of depreciation on each barn at 5%. Learn the cost of insurance and repairs for the year. Include the interest, depreciation, insurance, and repairs in one sum called the "Annual Cost." Divide this annual cost by the number of animals sheltered and learn the cost of sheltering one head on each of the farms. Where different animals are kept, they may be reduced to a comparative basis by estimating the weight and considering 1000 pounds as a unit.

AGRI-BUSINESS AND MARKETING

STATE SCHEME

Own building for Agmark Laboratories, Strengthening of Agmark Laboratories and Provision of Computers (₹3.07 crores): There are 30 State Agmark Grading Laboratories, one Principal Laboratory and 15 Agricultural marketing centres functioning in the State. It is proposed to construct building for Agmark Labs as well as modernize the equipments during the Tenth Five Year Plan for which a sum of ₹3.07 crores is provided.

Schemes Proposed to be Implemented Availing Financial Assistance from the other Financial Institutions

Provision of Infrastructure Facilities (₹ 100 crores)

i. *Creation of infrastructure facilities:* There is a need to provide infrastructure facilities like transaction shed, drying yard, farmers rest sheds, sanitation facilities, drinking water supply electronic weighing scales, cleaning and washing facility, moisture meters, scientific instruments for grading, rural godowns etc., for which an amount of ₹100 crores is provided.

ii. *Revamping Tamil Nadu Agricultural Marketing Board:* There is an imperative need to revitalize the Board activities by making this body more responsive and independent decision- making body. It should be given administrative and financial powers.

iii. *Revitalising the market committees:* At present the Act provides for nomination of members to the Market Committees. To make it more accountable and democratic, it is necessary that members be elected so that there is a sense of belonging. Secondly, there is lack of professionalism in the staff of the Market Committees. Their mind set is more of regulation than market oriented approach. Training to the existing staff and inclusion of professionals is a must.

iv. *Provision of infrastructure facilities in Post Harvest Centres:* At present the post harvest centres in the Regulated markets conduct training programmes to the farmers and exhibitions on various post harvest practices. The regulated markets should expand their activities through provision of whole storage CA, MA, Retardation and ripening facilities for the agricultural produce.

v. *Agricultural Marketing Extension:* Strong network of marketing extension is very much necessary at block level to effectively advise farmers on various aspects of marketing, advise on product planning, marketing information, and securing market for farmers, and advise on improved market practices and advise on post harvest management practices.

To strengthen the regulated market it is proposed to provide infrastructure facilities like transaction shed, drying yard, farmers resting shed, sanitation facilities, drinking water supply, electronic weighing scales, cleaning and washing facility, moisture meters, scientific instruments for grading, rural godowns etc. at a cost of ₹100 crores with financial assistance from Marketing Committee, NABARD, GOI.

Provision of infrastructure facilities in Post Harvest Centres-Cold Storage (₹ 200 crores- National Horticultural Board/Private Sector)

It is estimated that around 30% of the horticulture produce is wasted due to inadequate cold chain facility and appropriate technology for the preservation of horticultural produce. At present the combined cold storage capacity of 133 units in Tamil Nadu is around 1 LMT. The existing capacity is not sufficient to store horticultural, dairy, and marine products. In order to meet the demand it is proposed to establish an additional 1 LMT capacity of cold storage facility, at an estimated cost of ₹200 crore. Taking into consideration the initial capital investment, high recurring cost and low capacity utilisation and project failure, it has been proposed to attract investment by private sector through measures like financial incentives in addition to NHB's subsidy and power tariff concession.

Agricultural Marketing Extension/ Packaging Training (₹ 42.95 crores)- Tamil Nadu State Agricultural Marketing Board (TNSAMB)

Agricultural farmers require advice on various aspects of marketing like selection of the crops to be grown with marketability in mind, current price, market arrival and forecasting of market trends, and on post harvest management practices.

Agro-processing Food Park (₹ 10 crores)- Private Sector

It is estimated that about 10-20% of the food grain production and 30-35% of fruits and vegetables are wasted at various stages from picking to consumption. This indicates the need to establish food-processing industries to preserve and minimize the wastage. To encourage private sector to establish agro-processing industries, the Government of India, Ministry of Food Processing Industries has formulated two models of Food Park for Food Processing Industries apart from other incentives. It is proposed to set up a Food Park at a cost of ₹ 10 crores either in private sector or by Market Committees.

Setting up of Terminal Markets and Collection Centres (₹ 160 crores)

Private sector or growers association in partnership with private sector can organise Terminal Markets for specified products with backward integration with collection centres (value addition centres). Fifteen to twenty collection centres established nearer to the production area, can feed one Terminal Market. Two such a Terminal Markets will be established one each at Chennai and at Coimbatore with NDDB assistance.

Mega Markets (₹ 10 crores)

A mega market centre can cater to the needs of wholesale dealers, exporters and food processing industries. The availability of horticultural products in large scale in one place would promote exports and provide single sourcing to agri processing industries. It is proposed to establish a mega market for vegetables at Oddanchatram in Dindigul District at a cost of ₹ 2.59 crore. Depending on the success of this project, mega markets could be developed at Mettupalayam in Coimbatore District and Athur in Salam District either through Market Committees or private sector.

Agro Processing Industries (₹ 10 crores)

Food Processing Industries provide the crucial farm - industry linkages, which help to add value to the produce, generate employment opportunities and increase the net income to the farmers. The Market Committees will establish these through private sector in 10 places with an incentive of 20% of the capital cost. The project cost is estimated to be ₹ 10 crores. Packaging fresh fruits, vegetables and other farm produce is an important process as it reduces post harvest losses, increases the income of the farmers and ensures clean and hygienic farm produce to the consumers. It is proposed to create awareness and impart training on farm produce packaging to the technical officers of Agriculture, Horticulture, Agricultural Marketing and Agri Business departments during the Tenth Plan period. It is proposed to train 600 officials and 400 farmers at a total estimated cost of ₹25 lakhs. It is also proposed to conduct studies for developing low cost packaging techniques and material for some specific commodities at a cost of ₹0.35 crore.

Agri Export Zone (₹ 150 crores) –GOI

The most critical factor to meet the challenges of export will be enhancing exporting capability of our State in a highly competitive environment. To promote export of agri products, Government of India, Ministry of Commerce has announced in 2001 - Exim Policy, a scheme of establishing Agri Export Zone (AEZ). In AEZ, institutional and physical infrastructure would be created as per the needs of the specific commodity. Steps have been taken to establish AEZ for cut flowers, mangoes, bananas, medicinal plants and vegetables. To promote agricultural exports, it is proposed to educate and train the growers of identified crops in producing, grading and packing for international market, and in establishment of analytical laboratories, setting up of Export promotion cell in Agri Business department to disseminate information and harmonization of standards of Indian products with international standards (Codex).

MARKETING EXTENSION

Network is proposed to be formed integrating with the extension network already available with Agriculture Department. Officers of Agriculture, Horticulture and Agricultural Marketing departments will be given training on various aspects of Agricultural Production and Marketing for the purpose of carrying out extension works effectively and efficiently.

A sum of ₹65 lakhs is proposed during the Tenth Five Year Plan. Further, the Government will also facilitate private sector to carryout extension work for the quick reach of farm information to the farmers.

To improve market practices and also sale of specific agricultural products, two products' specific market Complexes *viz*- Turmeric Market Complex at Erode at a total cost of ₹32.30 crores, and Jaggery Market Complex at Trichy at a total cost of ₹10 crores will be established. These market complexes will be established incurring expenditure initially from Market Committees and financial institutions like NABARD and then recovered from the traders in instalments.

ALTERNATIVE MARKETING FORMS

Role of Government in managing markets is on decline worldwide. It is not easy to bring major changes in the traditional marketing system. The only way to modernize marketing is to promote alternative marketing system and that may operate parallel to and in addition to present marketing system. The purpose of the proposed alternative marketing is to promote modern trade practices, which in turn will pave way for transperancy and effiecncy in market.

Even though, the various forms of alternate marketing like:

- Direct marketing,
- Marketing through farmers interest group,
- Setting up of terminal markets,
- Forward and future market,
- e-commerce,
- Setting up of mega markets,
- Negotiable warehouse receipt system etc. have been suggested by Expert Committee on Agricultural Marketing headed by Shankarlal Guru, three important marketing methods could be considered in the State *viz.*, Terminal Market, Mega Market and Direct Market.

AGRO PROCESSING INDUSTRIES

Food Processing Industries provide the crucial farm - industry linkages which helps to accelerate overall agricultural development, adding value to the produce, generating employment opportunities and increasing the net income to the farmers. In India, only 2% of the total horticultural produce is processed. In countries like Brazil, it is in the range of 70%. Considering the rising demand for good quality products, there is an urgent need to enhance capacities for value added and processed products. At present, value addition is estimated at 7% of the total production within next 5 years.

There is a need to increase value addition to 20% and processing at 7%. To increase our country's share in the world trade of agri products which stands at less than 1% at present, the most critical factor in the highly competitive market environment is quality processed products. In Tamil Nadu, food processing in the form of drying, vegetable oil, grain processing, sugar breweries are in existence for a quite long period. Lack of adequate infrastructure facilities like storage, processing, marketing besides technical know-how have been the major constraints affecting the growth of the industry. Tamil Nadu with a coastline of 922 km and surface boundary of 1200 km with tropical and sub-tropical climate is ideally suited for the production of a host of agricultural, horticultural, acquacultural and animal husbandry produces. Thus there is vast scope for setting up of Food Processing Industries in Tamil Nadu.

The growth of food processing industry, which is included in the priority lending sector, will bring immense benefit to the economy. Economic liberalization and raising consumer prosperity is opening up new vistas in food processing sector. Tamil Nadu produces 103 lakh metric tonnes of food grains, 22 lakh metric tonnes of oil seeds, 37 lakh metric tonnes of sugarcane, 53.89 lakh metric tonnes of fruits, 54.65 lakh metric tonnes of vegetables, 2.95 lakh metric tonnes of spices and 6.94 lakh metric tonnes of plantation crops. 10 to 20% of food grain production is lost every year both at pre and post harvest stages. Similarly 30 to 35% of fruits and vegetables produced are also wasted at various stages from picking to final consumption. This only indicates the immense potential and scope that exists for setting up of food processing industries so that the available raw materials may be put into the most judicious use. This also indicates the scope that exists for adoption of post harvest modern techniques ensuring preservation and minimizing wastages of agricultural produce.

IMPROVEMENTS IN THE PRODUCTION AND MARKETING

Marketing projects were limited to products which were subject to common price and market regimes but, as the range of commodities within the price policy expanded, this was no obstacle. Article 12 provided an interpretation of the four categories listed in Article 11 and it is interesting, even at such an early stage in the history of the CAP, that 'the adaptation and guidance of production' meant 'the quantitative adaptation of production to outlets' and 'improvements in the quality of the products'. During the 1960s—and indeed for a large part of the 1970s—Reg. 17/64 provided the main source of funds for structural reform. The range of activities covered by the Regulation was very wide, encompassing production and marketing projects, and those which were a mixture of the two. Over the lifetime of the Regulation, projects to improve production structures received 53 per cent of the funding, marketing structures 41 per cent, and mixed projects 6 per cent.

The Community contribution to the investment was initially very low at 25 per cent for all projects but, in 1973, this was raised to 45 per cent of eligible expenditure in projects related to production, with the beneficiary only having to provide a minimum of 20 per cent of the cost (originally a minimum of 30 per cent). Thus, right from the beginning, the Community element in structural legislation was low, unlike the price and market policy where, by the end of the 1960s, FEOGA-Guarantee was providing 100 per cent of the support. However, it should be noted that, for structural measures generally, the tendency over time has been for the Community contribution to rise, particularly in the poorer regions or in the poorer Member States.

During its lifetime, Reg. 17/64 was extremely popular and far more projects were submitted for funding each year than were ultimately aided. The sectoral distribution of aid, 1964-1979. Three sectors —milk, wine, and meat—accounted for one-third of the aid; land improvement and hydraulic works accounted for about another third; and all other purposes the remainder.

The result was that the better-off farmers and the more prosperous farming regions were benefiting disproportionately from all aspects of the CAP. An interdepartmental group within the Commission reported *inter alia* on this issue, and its findings in relation to the regional impact of Dir. 72/159 have already been discussed. As to the regional impact of Reg. 17/64, if it had been operating as originally intended, each Community Programme would have had specified zones of principal action within which investment aid would have been concentrated. The Commission had estimated that, as a general rule, these zones would have represented about one-third of the designated area or other appropriate measure, such as volume of production.

In reality, in Reg. 17/64, there was no provision for regional differentiation of aid, rather the funds available were allocated to the Member States according to a prearranged system of quotas. Thus, prior to Enlargement, Italy received annually an allocation of just under 34 per cent of available credits, Germany 28 per cent, France 22 per cent, and so forth. These percentages were adjusted downwards to accommodate the three new Member States after 1972.

The interdepartmental group was interested to examine how these national quotas had operated in practice and to compare them with the percentage distribution among the Member States of: (a) the agricultural labour force; (b) final agricultural production; (c) the simple average of (a) and (b), which in effect gave production weighted by the inverse of agricultural labour productivity; and (d) agricultural area. In no case did the quota exactly match any of these measures, although it came reasonably close to (c). For some countries the divergence was considerable—which was hardly surprising, as the quotas had been devised more on a basis of horse-trading than on any economic or social indicator.

Within some of the Member States, the group discovered that the quotas were apparently being divided by region, according to the size of the agricultural area. This meant that 'some of the relatively extensive and well-structured agricultural regions of the community (such as Schleswig-Holstein, Niedersachsen, Scotland, Emilia-Romagna, Toscana and Lazio) receive much higher expenditure than agricultural employment or structural criteria would suggest'. The group also commented on the considerable difference between expenditure commitments for Italy and actual payments: in 1964-72 it received about one-third of total commitments but only half that amount had been paid by 1974. Thus, even at this early stage, administrative difficulties were preventing Italy from gaining access to structural funds—a pattern that was to become familiar as the years went by.

It is of some interest to examine whether or not the perceived lack of balance between the need for structural improvement in poorer regions and the distribution of total payments from Reg. 17/64 persisted over the lifetime of the scheme. In 1989, the Commission published a cumulative regional breakdown of the aid granted. While Italy received the largest payments, and while the South and the islands received a considerable share of the total, Emilia-Romagna remained a main beneficiary, and Lazio received more than Sicily.

Germany, which received the second highest payments, still had Niedersachsen as its second highest recipient, although Schleswig-Holstein did not maintain its earlier position. In the UK, Scotland remained by far the largest recipient. In Belgium, the south is poorer than the north but received less aid, while in Ireland two regions—South-East and South-West—dominated, although the West and North-West were much poorer.

Member State of the total number of projects financed. Most were already completed by the end of 1992. About 11.5 per cent had been abandoned, for whatever reason, by far the greatest number of these were in Italy. Despite its late arrival, the UK made very good use of Reg. 17/64. Member State of the aid granted, cumulatively to the end of 1992. This should be compared with the payments made. While the sources are not the same, which may have resulted in some discrepancies, in broad terms one can conclude that the account is nearly closed on this scheme. In Italy there is a considerable difference between aid granted and payments received, which presumably is a reflection of the large number of projects which were abandoned.

So the life of this extraordinary piece of legislation draws to a close. For so many years, it provided a link with the earliest days of the CAP and, in the distortions which so quickly developed in its operation, was a reminder of how soon in the life of the policy the politics of the CAP usurped the ideals of the original designers. However, some aspects of Reg. 17/64 have remained. In 1976, the interdepartmental group referred to earlier recommended that, when Reg. 17/64 ceased, its infrastructural aspects should be restricted geographically in any replacement legislation.

In a sense, this did occur with the measures which came into force in the late 1970s and early 1980s. However, it was not until the reorganization of the Structural Funds in the late 1980s that a more systematic approach to regional concentration of effort was introduced into the CAP. The aid to processing and marketing was carried forward with the enactment of Reg. 355/77.

THE ESTABLISHMENT OF PRODUCER GROUPS

One of the most obvious features of agricultural production in Western Europe (and in many other parts of the world) is its organizational structure, dominated by small and medium-size independent farms. Production in such a framework comes very close to the economist's definition of a perfect market. Farmers are price-takers: they are very rarely in a position to dominate the market, even if they are organized into a commodity or special interest group. In general, they produce a homogeneous product and therefore cannot benefit from higher prices achieved through product differentiation. They are in a weak bargaining position, whether selling a product for direct consumption or for processing, and this is particularly true when the product is perishable (as with milk) or perishable and seasonal (as with fruit and vegetables).

The poor structural organization of agriculture was even more pronounced in the 1960s in the new Community than it is in the 1990s. Farming was characterized by a very large number of exceedingly small producers, each with an extensive range of products. This situation was not conducive to orderly marketing, particularly at a time when consumer demand was changing rapidly due to the marked increase in living standards.

What was required was regular supply, known and constant quality, and stable prices. It was essential that producers respond and that they took greater responsibility for the marketing of their own products. Vague references were made by the Commission as far back as the 1959 and 1960 proposals on the CAP to its intention to encourage initiatives by farmers' organizations, which would lead to better information on markets, to improve their stability, and their response to consumer demand.

The Commission reasoned that, if producers of the same commodity came together in an alliance, it would enable them collectively to improve the quality of their product by enforcing grading standards, and the continuity of supply, by having centralized storage and processing facilities. The group would market the output of its members, thereby improving their bargaining power and, as a result, obtaining a higher unit price. This in turn would help to further the aims of Art. 39.1 of the Rome Treaty, in particular as regards incomes, market stability, and supply availability.

These thoughts were carried a stage further in Reg. 26/62 on the application of certain rules of competition to production of and trade in agricultural products. Under this Regulation, farmers and their organizations were, in general, excluded from the ordinary Community rules on competition, in so far as their activities were concerned with the production and sale of agricultural products, and the use of common facilities for the storage and processing of such products.

Although, theoretically, the case for collective action is strong, in practice a recurring difficulty with any sector dominated by small-scale independent producers is to persuade them to join a group which will have power to impose standards, control prices, and even production, and once they have joined to prevent them from breaking their contractual obligations to the group, thereby undermining its collective bargaining power.

One way in which the advantages of membership of a producer group can be enhanced is to give to such groups certain formal market intervention responsibilities. Prior to the establishment of the European Community, the Netherlands provided good examples of how such a system functioned, especially in the fruit and vegetable sector. The characteristics of the Dutch system were outlined in the 1960 proposals on the CAP, and certain features of that system found an echo in some of the commodity regimes. The first example came in Reg. 159/66 *concerning additional arrangements for the common organization of the market in the fruit and vegetable sector*. This Regulation completed the arrangements for fruit and vegetables which had been hanging fire since 1962. A novel feature of this regime is the important role played by producer groups which act as the agency for internal market support. For many commodities, provision is made in the relevant market regime for the withdrawal of produce from an oversupplied market, its storage, and disposal. These tasks are normally undertaken by an official agency or agencies established in each Member State. However, for a small number of products, market management is in the hands of the producers themselves, through producer groups which receive official recognition. Given the characteristics of fruit and vegetable production—a highly perishable product, many small-scale producers, local markets, seasonal production, volatility in supply and price-flexibility and speed of response are required if any form of market management is to be effective.

Apart from market support for certain commodities, more generally producer groups provide their members with facilities for handling and marketing the output of the relevant commodity. In their turn, the members are expected to market this total supply through the group (unless the group allows otherwise), and they must abide by the rules adopted by the group for the improvement of quality and the adaptation of supply to demand. The involvement of FEOGA-Guidance lies in the financial encouragement it provides for the establishment of producer groups. The first example of such aid appeared in Reg. 159/66 under which Member States were given the possibility of helping producer groups to get started, by providing them with a degressive aid during the three years following their establishment. This annual aid was not allowed to exceed 3, 2, and 1 per cent respectively of the value of the production marketed by the group. The Guidance Section reimbursed 50 per cent of the Member States' eligible expenditure under this scheme. Tentative references were made in Reg. 122/67 *on the common organization of the market in eggs* and in Reg. 123/67 *on the common organization of the market in poultrymeat* to the need to encourage efforts to improve the organization of production, processing, marketing, and quality through trade groups. However, in these commodity regimes, no financial support was offered to help launch such groups.

In the light of all this piecemeal and unco-ordinated interest in producer organizations in various commodities, it is not surprising that the Commission eventually proposed the establishment of a Community framework within which Member States could encourage the formation of producer groups in any commodity. The Commission was also prompted to do so because certain Member States were already promoting these groups and there was a danger that problems would arise at Community level through the use of different criteria and objectives in the various countries concerned.

Thus, in 1967, the Commission proposed a Regulation on producer groups and their unions in COM(67)68 final, which set out common rules on the type of aid which could be given by the Member States to encourage the formation and development of these groups, and on the conditions under which they could be officially recognized. Despite the Commission's evident interest in the creation of producer groups and despite the existing Community launching aids in the fruit and vegetable sector, the 1967 proposal did not make provision for Community aid. This was somewhat surprising and indeed was criticized by the European Parliament in its Resolution on the draft.

Although Mansholt always regarded producer groups as an extension of market organization rather than as a part of structural policy, his 1968 Plan did contain a discussion of the need to improve marketing, and one of the measures suggested to aid that process was the immediate adoption by the Council of the proposed Regulation on producer groups. The Council clearly did not heed this exhortation because, in its 1970 structural reform proposals, the Commission made a pointed reference to the thorough discussions which had taken place in the European Parliament and in the ESC on the 1967 producer group draft and stated that the Council had not even begun a discussion on it. The 1970 reform proposals included a considerably revised draft Regulation on producer groups—the most important change being the provision of Community aid to help launch the groups. In the explanatory note to the proposals, the Commission made it clear that it regarded the improvement of market structure as primarily the responsibility of the Member States. For that reason, it proposed that the Community should refund only 30 per cent of eligible national expenditure. The logic of this position is unclear, as it will be recalled that the Community was already refunding 50 per cent of eligible expenditure on producer groups in the fruit and vegetable sector.

In 1971, when the revised structural reform package was presented to the Council, the draft Regulation on producer groups appeared again—now in a third version. The changes made were generally of a technical nature but, importantly, the Community contribution to eligible expenditure by the Member States had been reduced to 25 per cent. It was all to no avail, as the draft Regulation was not passed by the Council.

In the *Resolution* on the new orientation of the Common Agricultural Policy in 1971, the Council finally agreed in principle to the introduction of a regime to encourage the formation of producer groups but, once again, it appeared that Community aid would not be offered and that the funding would be by the Member States. Despite this, however, the Council remained unable to agree to the proposal, although discussions dragged on into 1972. The only progress which was made was under Reg. 1696/71 *on the common organization of the market in hops*.

The hops producer groups are not involved in price support operations but are intended in a more general way to assist in the centralization of supply and in the adaptation of the product to market requirements. Member States were free to grant a launching aid and FEOGA-Guidance reimbursement was limited to 25 per cent of eligible expenditure. An extra feature of this regime was the option given to Member States to aid producer groups on a temporary basis to reorganize hop gardens and to switch to more suitable varieties. The form of aid was a payment per hectare replanted or reorganized, and FEOGA-Guidance refunded 50 per cent of the eligible expenditure of participating Member States.

Discussions resumed in 1976 on the introduction of a general producer group regime and, in the following year, the Commission put forward yet another revised proposal. This was so radically different from previous drafts as to be, in effect, a completely new proposal. In the explanatory memorandum, the Commission tried to explain why such a harmless piece of legislation was causing such difficulties.

Basically the problem was that certain Member States with well-developed production and marketing structures could not see the necessity for this legislation. If anything, the differences between the Member States had grown since the original 1967 proposal. The Commission, therefore, proposed that instead of a Regulation of universal application, there should be a measure which would apply only in those areas where the supply of agricultural produce was structurally very inadequate. Primarily this meant Italy and indeed the proposal referred solely to that country, with the provision that the Regulation could be extended to other Member States.

The Council had agreed in February 1977 that it would adopt the proposal by the end of June. It was in fact a full year later that Reg. 1360/78 *on producer groups and associations thereof* was enacted. It was almost totally unrecognizable from the draft and covered not just Italy but also Belgium, Southern France, and DOM (the French overseas departments). The commodity coverage differed by country, and in France by region. It was intended to be of assistance to farmers in regions and sectors where there were severe structural deficiencies, due to the small size of farms and poor organization. For a measure which was to last initially for five years only, at a total FEOGA cost of 24m. ua, an incredible amount of bureaucratic effort was involved. When Greece, Portugal, and Spain joined the Community, Reg. 1360/78 was extended to them, and to Ireland and the remainder of France in 1988, all for specified lists of commodities.

The only other commodity for which producer group launching aids are available is cotton under Reg. 389/82. Under the terms of Accession, when Greece joined the European Community, market support was extended to cotton and producer groups were seen as an important mechanism for the rationalization of production and the centralization of handling and marketing. For this reason, the Regulation covers not only launching aids but also investment aid to assist with structural improvements in harvesting, ginning, storage, and packaging. Producer groups in Italy and Spain are also eligible for aid under this regime. Having sketched the rather confused history of the development of legislation on producer groups, the remainder of this Section reports on the statistical evidence of the success or failure of the various structural measures to persuade farmers to come together to improve the organization of particular commodities. Data on structural aids are available for producer groups in fruit and vegetables, hops, cotton, and miscellaneous products under Reg. 1360/78.

In 1972, Reg. 159/66 was replaced by Reg. 1035/72, which consolidated all the existing Community legislation on fruit and vegetables. Arrangements for launching aids for producer groups were retained unchanged. The pattern which is so familiar from other structural aids is present here also: aid is concentrated in a very small group of countries and it is not necessarily related to the size of the sector. For instance, the very small number of producer groups set up under this legislation in Belgium and the Netherlands is indicative of the strength of the previously long-established marketing structures based largely on the cooperative movement. Therefore, there was little incentive or need to seek launching aids. In contrast, the relatively large number of launching aids in the UK is an indication of the previous absence of joint ventures by farmers rather than an indication of the size of the horticultural sector. The size of producer groups varies considerably from one Member State to another, as evidenced by the absence of any relationship between the number of groups and the level of investment.

Hops provide a real contrast to fruit and vegetables. This is a very small sector, highly localized in traditional areas of production where, of course, it can be quite a significant crop. Much of the output is grown on contract. The schemes of aid came to an end some years ago and 1985 was the last year for which data were available. The situation with regard to cotton is quite different. Greece is the largest producer and so far (at the time of writing) was the only Member State to have made use of the launching and investment aids, although the Commission anticipated that Spain, which is the only other significant producer, might well begin to show an interest in this measure.

In 1991, the Commission reported to the Council on the operation of Reg. 1360/78. By the end of 1990, the Member States had granted recognition to 560 groups. The discrepancy is explained by 'the slowness of the Member States' administrations in drawing up implementing rules and requesting refunds'. Despite the fact that Belgium was included in the scheme right from the start, no groups in the relevant production sectors (cereals, cattle, pigs, and Lucerne) had applied for recognition, even though the requirements had been made easier.

Indeed, the Commission stated that the Belgian authorities had reported 'that those concerned are uninterested in the very idea of producer groups and that, in the present circumstances; this is unlikely to change in the future'. For that reason, the Commission proposed that, when the life of the Regulation was prolonged, Belgium should be excluded. This was rejected!

IMPROVEMENTS IN PROCESSING AND MARKETING

The fact that the processing and marketing support activities of Reg. 17/64 were continued in Reg. 355/77 on common measures to improve the conditions under which agricultural products are processed and marketed. This Regulation was a long time in coming. The life of Reg. 17/64 was limited following the introduction of the financial Regulation 729/70, which laid down the basis for future assistance of a structural nature. If for no other reason than that of legal necessity, consideration had to be given to the replacement of Reg. 17/64 by some other measure in the sphere of marketing and processing.

More than that, however, there has always been a tendency to concentrate aid to agriculture in that sector of the industry which lies on the farm and to neglect the need to improve the handling of the products once they leave the farm. For many years, Reg. 17/64 was the only Community measure of any importance providing such post-farm aid, despite the fact that it was recognized that the agricultural sector was weak in the handling of produce.

In setting the new guidelines for the CAP in 1971, the Council had invited the Commission to continue to study the problems of marketing and processing, and to submit a proposal on their alleviation. In 1973, in the *Improvement* memorandum, the Commission announced its intention to make such a proposal but it was not until mid-1975 that it finally managed to submit this proposal to the Council in COM(75)431. The Commission reasoned that a more efficient processing and marketing sector would be in a position to pay better prices to producers, to diversify output, thereby stimulating demand, to concentrate more on exports, and (rather obscurely) to 'better handle products from remote areas of the Common Market'.

The proposal was for the establishment of multi-annual programmes, each of which would cover one or more agricultural products. Within these programmes the Community would provide aid for suitable projects intended to promote the rationalization and expansion of processing and marketing. Assistance was to take the form of a capital subvention of not more than 25 per cent of the value of the investment, with the beneficiary providing at least 50 per cent and the Member State participating in the funding as well. The estimated cost of the scheme was 400m. ua over a five-year period.

While largely in agreement with the intentions of the Commission, the ESC had a number of reservations.

For one, the Committee was of the opinion that the proposed Regulation was 'of direct concern to consumers, farm workers, and workers in the food industry and the distributive trade' but that insufficient regard was being paid to the interests of these parties.

The ESC also believed that the effectiveness of the measure could be improved by the formation of producer groups and urged the Council to adopt the long out standing draft Regulation on such groups. This point was echoed by the European Parliament.

Another point on which the two bodies agreed was their scepticism concerning the Community's ability to control the new measure in a manner which would avoid exacerbating the surplus problem. Both Parliament and ESC considered the proposed scheme as financially inadequate. The Parliament stated that 'the Commission's proposal represents a limited step which will result in a decrease in the total real amount of Community aid to be granted for the improvement of marketing and mixed production/marketing structures, and which will make no substantial contribution to reducing agricultural surpluses and limiting the need for intervention'.

The ESC commented that '80 million units of account per year is not enough to solve all the problems of improvement of agricultural product marketing and processing facilities, especially as costs will continue to be pushed up by inflation'. It was not until 1977 that a modified version of this proposal was passed by the Council and it is hard to believe that either the Parliament or the ESC would have regarded the changes as beneficial.

The programmes 'to develop or rationalize the treatment, processing or marketing of one or more agricultural products' had to be drawn up by each Member State and agreed by the Commission before applications for the funding of individual projects could be considered. These projects could be for a wide range of activities: the development of new outlets for agricultural products; lightening the burden on intervention by improvements to the structure of a market; assistance to regions having difficulty adjusting to some aspect of the CAP; improvement of marketing channels or rationalization of the processing of an agricultural product; improvement of quality or presentation... While the contribution from FEOGA-Guidance to eligible expenditure was fixed at 25 per cent, in later years this was modified in various countries and regions, *e.g.* in the Mediterranean, and the West of Ireland. Strange little schemes were added on from time to time as, for instance, a special measure to rationalize and improve slaughterhouses in Belgium.

The support arrangements for processing and marketing were overhauled in 1990, so as to bring them into line with the new arrangements for the Structural Funds.

However, in some Member States there have been shifts, with new regions coming to the fore, such as the North East in Ireland. There still was evidence that the poorest regions were not achieving the enhanced level of investment necessary for them to narrow the income gap. There were great differences by country, with one or two sectors in each dominating: meat in Belgium; fish in Denmark; cereals and wine in Germany; fruit and vegetables, and tobacco in Greece; fruit and vegetables, and meat in Spain; wine, and fruit and vegetables in France; meat in Ireland; fruit and vegetables, and wine in Italy; fruit and vegetables in Portugal; and fish and meat in the UK. Taking the Community as a whole, fruit and vegetables, meat, and wine projects dominated in terms of number, and together they also accounted for over 57 per cent of the funds.

As with Reg. 17/64, the outstanding feature of the table is the relatively large number of projects abandoned in Italy—over 38 per cent of the total. The figure for the UK—although much lower—is also rather high. The second part of the table gives the cumulative figures for funds committed and actually paid. The rate of reimbursement is an indication of the efficiency of the programme in the Member States, both in terms of the take-up of the investment opportunities and of the national bureaucracies. Five Member States had achieved a payment/commitment level in excess of 70 per cent; France and the UK achieved levels in excess of 60 per cent; Ireland and Italy in excess of 50 per cent. Greece, Portugal, and Spain were well below these levels. While the last two countries had only recently joined the Community and therefore could be expected to be somewhat slow the same cannot be said for Greece.

In terms of total funds, aid to marketing and processing is one of the biggest sources of structural investment in agriculture—just as Reg. 17/64 had been before it. Together with funds spent on improving the efficiency of farms, and LFAs, these three are the most significant programmes. While aid to LFAs has a large social element in it, in that it is intended to compensate for inherent locational handicap, the other major aids are intended to make lasting improvements to the economic framework within which farming takes place and agricultural produce reaches the market.

Despite their popularity, one must question to what extent they have been successful in raising levels of efficiency, so that one could see a time when the need for palliative measures under the price and market policy would be reduced or even eliminated. If such a time is not yet in sight, then one needs to question whether the commitment to and content of structural policy within the CAP are adequate.

CO-OPERATIVE FARMING

Factors like increasing pressure of population on land, uneconomic size of holding, primitive and unscientific methods of cultivation, inequitable distribution of land, poverty and ignorance of the peasantry, *etc.*, are the major impediments that stand in the way of implementation of the planned agricultural production in India. However, co-operative farming which implies pooling together of the scattered and uneconomic holdings of land and their joint management will go a long way towards the progress of agriculture. There are four types of co-operative farming societies which are described in the following paragraphs. They are:

- Better farming society,

- Tenant farming society,
- Joint farming society, and
- Collective farming society.

BETTER-FARMING SOCIETY

The better-farming society could be said to form the basis of the co-operative farming programme. The main object of it is to educate and to prepare the farmers to accept the new system of farming. For this, they organise demonstrations of improved methods of agriculture. Use of improved seeds, manures and implements is the most common activity undertaken by these societies. Besides this, a number of other activities such as disposal of farm produce at reasonable prices, purchase of occupational requisites, *etc.*, are also undertaken. Under this type of co-operative farming, the ownership and management of land rest with the individual.

CO-OPERATIVE TENANT FARMING SOCIETY

The co-operative tenant farming society provides its members with facilities such as finance, implements, seed, *etc.* The society owns land or gets it on lease, but it does not undertake farming. Land is divided into blocks and each block is given on rent to a cultivator who cultivates according to the plan laid down by the society.

CO-OPERATIVE JOINT FARMING SOCIETY

The joint-farming societies are suitable to solve the problem of fragmentation of land and the cultivation of uneconomic holdings. The land of small owners is pooled together increasing thereby the size of the unit of cultivation.

The members of the society work jointly on the pooled land according to the programme of the society. The cultivators who work on the farm receive wages for their labour. As against the proprietorship of land, the owner cultivators get dividend or rent in proportion to the value of the land. The common functions of these types of societies are planning of crop programmes, joint purchase of farm requisites, joint cultivation, raising of funds for the improvement of land and joint sale of farm produce. The small owners of land are encouraged to pool their land so as to form a large unit of cultivation. The society can also purchase or take on lease land for cultivation. Out of the proceeds received from the disposal of the produce, all the expenses of cultivation including payment for the use of land, wages and cost of management are met first. Provision for interest on borrowings, depreciation of wasting assets, previous losses and for reserves and other funds is also made. The residue is then shared by members in proportion to the wages earned by each after utilising a part thereof towards the payment of bonus to the salaried staff.

CO-OPERATIVE COLLECTIVE FARMING SOCIETY

The society owns land or gets it on lease and it is collectively cultivated by its members. Most of these societies are organised on Government waste lands. Members get wages for their work and in the case of profits a bonus is paid in proportion to their wages. No dividend is paid on the share capital. The members of the collective farming society do not have any ownership or proprietary rights in the land.

The organisational position of co-operative farming in Sangli district as on 30th June 1963 was as follows: —

Of the 16 co-operative farming societies 10 were co-operative collective farming and six were co-operative joint farming societies. Eight collective farming societies were organised by backward class people. Out of the 16 societies nine collective farming societies and two joint farming societies were actually working while four were newly registered and one was defunct.

As the co-operative farming societies depend mainly on the Government lands which are leased out on short-term basis and which are generally on hilly slopes where considerable labour and money are required for cultivation, the progress of the movement is, slow. Only one co-operative farming society at Salagare in Miraj taluka had land on a permanent basis.

FARMING SYSTEMS IN INDIA

Farming Systems in India are strategically utilized, according to the locations where they are most suitable. The farming systems that significantly contribute to the domestic GDP of India are subsistence farming, organic farming, and industrial farming. Regions throughout India differ in types of farming they use; some are based on horticulture, ley farming, agroforestry, and many more. Due to India's geographical location, certain parts experience different climates, thus affecting each region's agricultural productivity differently. India is very dependent on its monsoon-based per for large India agriculture has an extensive background which goes back to at least 10 thousand years. Currently the country holds the second position in agricultural production in the world. In 2007, agriculture and other indu made up more than 16per cent of India's GDP.

Despite the steady decline in agriculture's contribution to the country's GDP, India agriculture is the biggest industry in the country and plays a key role in the socioeconomic growth of the country. India is the second biggest producer of wheat, rice, cotton, sugarcane, silk, groundnuts, and dozens more. It is also the second biggest harvester of vegetables and fruit, representing 8.6per cent and 10.9per cent of overall production, respectively. The major fruits produced by India are mangoes, papayas, sapota, and bananas. India also has the biggest number of livestock in the world, holding 281 million. In 2008, the country housed the second largest number of cattle in the world with 175 million.

BRIEF HISTORY

Agriculture in India began in about 9000 BCE when Indians learned to cultivate plants and domesticate crops such as wheat, barley, and jujube and animals such as goat and sheep. Cotton was cultivated around the 5th millennium BCE and became a well developed industry with the continuation of many methods of cotton spinning and fabrication into the modern industrialization of India. By 4500 BCE, the development of irrigation made the Indian continent prosper. As a result, Indian civilization grew, leading to more planned settlements that made use of drainage and sewers and introduced the development of storage systems such as artificial reservoirs and canals, making irrigation much more sophisticated.

Around 200 CE, the Tamil people, who were an ethnic group native to Tamil Nadu, India, cultivated a vide variety of crops such as rice, sugarcane, black pepper, various grains,coconuts, beans, *etc.* The spice trade gained momentum in Indian agriculture as spices native to India such as cinnamon and black pepper were being shipped to the Mediterranean. Crystallized sugar was discovered around 320 CE and soon after technology for sugar-refining developed.

As technologies were being further advanced, irrigation systems were introduced which gave rise to economic growth. The landscape was divided into agricultural 'zones', each producing rice, wheat or millets. Rice production was dominant in Gujarat while wheat dominated north and central India. Other crops to be cultivated were introduced during this period such as tobacco, tea, coffee, pineapples and papaya.

From the 18th to the 20th century, India was under British control. The combination of various factors, especially the war in 1918(world war 1) and being under British rule stagnated India's agricultural economy. Very few of India's commercial crops such as cotton, indigo, opium, and rice made it to the global market. Food and non-food outputs were declining as population was increasing, leaving India in an acute crisis. As the market for irrigation developed, community effort and private investment soared. This eventually helped get India's agriculture back on its feet.

Since, India's independence, food and cash crop supply has greatly improved with the establishment of special programmes such as The Grow More Food Campaign in the 1940s and the Integrated Production Programme in the 1950s. Land reclamation and development, mechanisation, electrification, and the use of chemicals soon followed agricultural development. Before India's agricultural and economic fall under British control, India's entire agriculture was practised organically; materials like fertilizers and pesticides were obtained from plant and animal products. Organic farming shifted to chemical farming in the 1960s when the Green Revolution became the government's most important programme for sustaining a rich and stable agricultural economy.

India has become one of the largest producers of wheat, edible oil, potatoes, spices, rubber, tea, fishing, fruits, and vegetables in the world. Between 2003 and 2004, agriculture accounted for 22per cent of India's GDP and employed 58per cent of the country's workforce and continues to hold these statistics today

CLIMATE EFFECT ON FARMING SYSTEMS

Each region in India has a specific soil and climate that is only suitable for certain types of farming. Many regions on the western side of India experience less than 50 cm of rain annually, so the farming systems are restricted to cultivate crops that can withstand drought conditions and farmers are usually restricted to single cropping. Gujarat, Rajasthan, South Punjab, and northern Maharashtra all experience this climate and each region grows such suitable crops like jowar, bajra, and peas. On the contrary, the eastern side of India has an average of 100–200 cm of rainfall annually without irrigation, so these regions have the ability to double crop. West Coast, West Bengal, parts ofBihar, U.P. and Assam are all associated with this climate and they grow crops such as rice, sugarcane, jute, and many more.

There are three different types of crops that are cultivated throughout India. Each type is grown in a different season depending on their compatibility with certain weather. Kharif crops are grown at the start of the monsoon until the beginning of the winter, relatively from June to November. Examples of such crops are rice, corn, millets, groundnut, moong, and urad.

IRRIGATION FARMING

Irrigation farming is when crops are grown with the help of irrigation systems by supplying water to land through rivers, reservoirs, tanks, and wells. Over the last century, the population of India has tripled. With a growing population and increasing demand for food, the necessity of water for agricultural productivity is crucial. India faces the daunting task of increasing its food production by over 50 percent in the next two decades, and reaching towards the goal of sustainable agriculture requires a crucial role of water. Empirical evidence suggests that the increase in agricultural production in India is mostly due to irrigation; close to three fifths of India's grain harvest comes from irrigated land. The land area under irrigation expanded from 22.6 million hectares in FY 1950 to 59 million hectares in FY 1990. The main strategy for these irrigation systems focuses on public investments in surface systems, such as large dams, long canals, and other large-scale works that require large amounts of capital.Between 1951 and 1990, nearly 1,350 large- and medium-sized irrigation works were started, and about 850 were completed.

PROBLEMS FROM IRRIGATION

Because funds and technical expertise were in short supply, many projects moved forward at a slow pace, including The Indira Gandhi Canal project. The central government's transfer of huge amounts of water from Punjab to Haryana a d Rajasthan contributed to the civil unrest in Punjab during the 1980s and early 1990s. Problems also have arisen as ground water supplies used for irrigation face depletion. Drawing water off from one area to irrigate another often leads to increased salinity receiving water through irrigation are poorly managed or inadequately designed; the result often is too much water and water-logged fields incapable of production.

Geography of Irrigation in India

Irrigation farming is very important for crop cultivation in regions of seasonal or low rainfall. Western U.P., Punjab, Haryana, parts of Bihar, Orissa, A.P., Tamil Nadu, Karnataka, and other regions thrive on irrigation and generally practice multiple or double cropping. With irrigation, a large variety of crops can be produced such as rice, sugarcane, wheat, and tobacco.

SHIFTING CULTIVATION

Shifting cultivation is a type of subsistence farming where a plot of land is cultivated for a few years until the crop yield declines due to soil exhaustion and the effects of pests and weeds. Once crop yield has stagnated, the plot of land is deserted and the ground is cleared by slash and burn methods, allowing the land to replenish. This type of cultivation is predominant in the eastern and north-eastern regions on hill slopes and in forest areas such as Assam, Meghalaya, Nagaland, Manipur, Tripura, Mizoram, Arunachal Pradesh, Madhya Pradesh, Orissa, and Andhra Pradesh. Crops such as rain fed rice, corn, buck wheat, small millets, root crops, and vegetables are grown in this system. Eighty-five percent of the total cultivation in northeast India is by shifting cultivation. Due to increasing requirement for cultivation of land, the cycle of cultivation followed by leaving land fallow has reduced from 25–30 years to 2–3 years. This significant drop in uncultivated land does not give the land enough time to return to its natural condition. Because of this, the resilience of the ecosystem has broken down and the land is increasingly deteriorating.

Shifting Cultivation in Orissa

Orissa accounts for the largest area under shifting cultivation in India. Shifting cultivation is locally known as the podu cultivation. More than 30,000 km^2 of land (about 1/5 land surface of Orissa) is under such cultivation. Shifting cultivation is prevalent in Kalahandi, Koraput, Phulbani and other southern and western districts. Tribal communities such asKondha, Kutia Kondha, Dongaria Kondha, Lanjia Sauras, and Paraja are all involved in this practice. Many festivals and other such rituals revolve around the podu fields, because the tribals view podu cultivation as more than just a means of their livelihood, they view it as a way of life.

In the first year of podu cultivation, tribals sow kandlan (variety of arhar dal). Sowing means spraying the seeds and is used at pre-monsoon time and the area is adequately protected. Yield differs from area to area depending on local climatic factors. After harvest, the land is left fallow. During the pre-monsoon, varieties of rice, corn and ginger are also sown. Generally, after the third year, the tribals abandon this land and shift to new land. On the abandoned land, natural regeneration starts from the available root stocks and seed bank. Bamboo comes up naturally; along with many other climbers that regenerate. Generally, this land is not cultivated for the next 10 years.

Impacts of Shifting Cultivation

Frequent shifting from one land to the other has affected the ecology of these regions. The area under natural forest has declined; the fragmentation of habitat, local disappearance of native species and invasion by exotic weeds and other plants are some of the other ecological consequences of shifting agriculture. Areas that have a fallow cycle of 5 to 10 years are more vulnerable to weed invasion compared to 15-year cycles, which have more soil nutrients, larger variety of species, and higher agronomic yield.

COMMERCIAL AGRICULTURE

In a commercial based agriculture, crops are raised in large scale plantations or estates and shipped off to other countries for money. These systems are common in sparsely populated areas such as Gujarat, Punjab, Haryana, and Maharashtra. Wheat, cotton, sugarcane, and corn are all examples of crops grown commercially.

Types of Commercial Agriculture

Intensive commercial farming: This is a system of agriculture in which relatively large amounts of capital or labour are applied to relatively smaller areas of land. It is usually practiced where the population pressure is reducing the size of landholdings. West Bengal practices intensive commercial farming.

Extensive commercial farming: This is a system of agriculture in which relatively small amounts of capital or labour investment are applied to relatively large areas of land. At times, the land is left fallow to regain its fertility. It is mostly mechanized because of the cost and availability of labour.

It usually occurs at the margin of the agricultural system, at a great distance from market or on poor land of limited potential and is usually practiced in the tarai regions of southern Nepal. Crops grown are sugarcane, rice and wheat.

Plantation agriculture: Plantation is a large farm or estate usually in a tropical or sub-tropical country where crops are grown for sale in distant markets rather than local consumption.

Commercial grain farming: This type of farming is a response to farm mechanization and it is the major type of activity in the areas of low rainfall and low density of population where extensive farming is practiced. Crops are prone to the vagaries of weather and droughts and mono culture of wheat is the general practice.

LEY FARMING

With increases in both human and animal populations in the Indian arid zone, the demand for grain, fodder, and fuel wood is increasing. Agricultural production in this region is low due to the low and uneven distribution of rainfall (100–400 mm yr"1) and the low availability of essential mineral nutrients.

These demands can be met only by increasing production levels of these Aridisols through adoption of farming technologies that improve physical properties as well as biological processes of these soils. Alternate farming systems are being sought for higher sustainable crop production at low input levels and to protect the soils from further degradation.

In India's drylands, ley farming is used as a way to restore soil fertility. It involves rotations of grasses and food grains in a specific area. It is now being promoted even more to encourage organic farming, especially in the drylands. Ley farming acts as insurance against crop failures by frequent droughts. Structurally related physical properties and biological processes of soil often change when different cropping systems, tillage, or management practices are used. Soil fertility can be increased and maintained by enhancing the natural soil biological processes. Farming provides balanced nutrition for sustainable production through continuous turnover of organic matter in the soil.

PLANTATION FARMING

This extensive commercial system is characterized by cultivation of a single cash crop in plantations of estates on a large scale. Because it is a capital centered system, it is important to be technically advanced and have efficient methods of cultivation and tools including fertilizers and irrigation and transport facilities. Examples of this type of farming are the tea plantations in Assam and West Bengal, the coffee plantations in Karnataka, Kerala, and Tamil Nadu, and the rubber plantations in Kerala and Maharashtra.

Forestry

In contrast to a naturally regenerated forest, tree plantations are typically grown as even-aged monocultures, primarily for timber production. These plantations are also likely to contain tree species that would not naturally grow in the area. They may include unconventional types of trees such as hybrids, and genetically modified trees are likely to be used in the future. Plantation owners will grow trees that are best suited to industrial applications such as pine, spruce, and eucalyptus due to their fast growth rate, tolerance of rich or degraded agricultural land, and potential to produce large quantities of raw material for industrial use. Plantations are always young forests in ecological terms; this means that these forests don't contain the type of growth, soil or wildlife that is typical of old-growth natural ecosystems in a forest.

The replacement of natural forest with tree plantations has also caused social problems. In some countries, there is little concern or regard for the rights of the local people when replacing natural forests with plantations. Because these plantations are made solely for the production of one material, there is a much smaller range of services for the local people. India has taken measures to avoid this by limiting the amount of land that can be owned by someone. As a result, smaller plantations are owned by local farmers who then sell the wood to larger companies.

Teak and Bamboo

Teak and bamboo plantations in India are a good alternative crop solution to farmers of central India, where conventional farming is popular.

Due to rising input costs of farming many farmers have grown teak and bamboo plantations because they only require water during the first two years. Bamboo, once planted, provides the farmer with output for 50 years until it flowers. Production of these two trees positively impacts and contributes to the climate change problem in India.

CROP ROTATION

Crop rotation can be classified as a type of subsistence farming if there is an individual or communal farmer doing the labour and if the yield is solely for their own consumption. It is characterised by different crops being alternately grown on the same land in a specific order to have more effective control of weeds, pests, diseases, and more economical utilization of soil fertility.

In India, leguminous crops are grown alternately with wheat, barley, and mustard. An ideal cropping system should use natural resources efficiently, provide stable and high returns, and avoid environmental damage.

Many farmers in India utilise the crop rotation system to improve or maintain soil fertility, check erosion, reduce the build-up of pests, spread the workload on family labour, mitigate the risk of weather changes, become less reliant on agricultural chemicals, and increase the net profit.

Different Sequences of Crop Rotation

Table. Rotation of Two Crops within a Year.

Year 1	Wheat
Year 2	Barley
Year 3	Wheat again

Table. Three Crop Rotation.

Year 1	Wheat
Year 2	Barley
Year 3	Mustard
Year 4	Wheat again

Pearl Millet

Pearl millet crop is mostly grown as a rainfed monsoon crop during kharif (June–July to September–November) and also as an irrigated hot weather (February–June) crop in central and south India. Pearl millet is often grown in rotation with sorghum, groundnut, cotton, foxtail millet, finger millet (ragi), castor, and sometimes, in the south India, with rice.

On the red and iron rich soils of Karnataka, pearl millet and ragi rotation is practised although pearl millet isn't always grown annually. Cluster bean – Pearl millet crop sequence with crop residue incorporation has significantly increased the productivity in the arid zone of Western Rajasthan where Fallow – Pearl millet/Pearl millet after Pearl millet crop sequence is practised. In Punjab, the dryland rotation may be small grain-millet-fallow. In irrigated lands, pearl millet is rotated with chickpea, fodder sorghum, and wheat. In the dry and light soils of Rajasthan, southern Punjab and Haryana, and northern Gujarat, pearl millet is most often rotated with a pulse-like moth or mungbean, or is followed by fallow, sesame, potato, mustard, moth bean, and guar. Sesame crop may be low-yielding and may be replaced by castor or groundnut.

DAIRY FARMING

In 2001 India became the world leader in milk production with a production volume of 84 million tons. India has about three times as many dairy animals as the USA, which produces around 75 million tons. Dairy Farming is generally a type of subsistence farming system in India, especially in Haryana, the major producer of milk in the country. More than 40per cent of Indian farming households are engaged in milk production because it is a livestock enterprise in which they can engage with relative ease to improve their livelihoods. Regular milk sales allow them to move from subsistence to earning a market-based income. The structure of the livestock industry is globally changing and putting poorer livestock producers in danger because they will be crowded out and left behind. More than 40 million households in India are at least partially dependent on milk production, and developments in the dairy sector will have important repercussions on their livelihoods and on rural poverty levels. Haryana was chosen to assess possible developments in the Indian dairy sector and to broadly identify areas of interventions that favour small-scale dairy producers. A methodology developed by the International Farm Comparison Network (IFCN) examined impacts of change on milk prices, farm management and other market factors that affect the small-scale milk production systems, the whole farm and related household income.

CO-OPERATIVE FARMING

Co-operative farming refers to pooling of farming resources such as fertilizers, pesticides, farming equipments such as tractors. It however generally excludes pooling of land unlike in collective farming where pooling of land is also done. Co-operative farming is a relatively new system in India. Its goal is to bring together all of the land resources of farmers in such an organised and united way so that they will be collectively in a position to grow crops on every bit of land to the best of the fertility of the land. This system has become an essential feature of India's Five Year Plans. There is an immense scope for co-operative farming in India although the movement is as yet in it infancy. The progress of co-operative financing in India has been very slow. The reasons are fear of unemployment, attachment to land, lack of proper propaganda renunciation of membership by farmers and existence of fake societies.

2

Cooperative Credit Management in Agriculture

COOPERATIVE CREDIT MOVEMENT OVER THE YEARS

Rural credit cooperatives in India were originally envisaged as a mechanism for pooling the resources of people with small means and providing them with access to different financial services. Democratic in features, the movement was also an effective instrument for development of degraded waste lands, increasing productivity, providing food security, generating employment opportunities in rural areas and ensuring social and economic justice to the poor and vulnerable.

The history of the cooperative credit movement in India can be divided in four phases. In the First Phase (1900-30), the Cooperative Societies Act was passed (1904) and "cooperation" became a provincial subject by 1919. The major development during the Second Phase (1930-50) was the pioneering role played by RBI in guiding and supporting the cooperatives. However, even during this phase, signs of sickness in the Indian rural cooperative movement were becoming evident. The 1945 Cooperative Planning Committee had discerned these signs in the movement, finding that a large number of cooperatives were "saddled with the problem of frozen assets because of heavy overdues in repayment." Even so, also in the Third Phase (1950-90), the way forward was seen to lie in cooperative credit societies. The All India Rural Credit Survey was set up which not only recommended state partnership in terms of equity but also partnership in terms of governance and management. NABARD was also created during this phase. The Fourth Phase from 1990s onwards saw an increasing realization of the disruptive effects of intrusive state patronage and politicisation of the cooperatives, especially financial cooperatives, which resulted in poor governance and management and the consequent impairment of their financial health. A number of Committees were therefore set up to suggest reforms in the sector.

OUTREACH OF THE COOPERATIVE CREDIT SECTOR

Of high relevance for financial inclusion of lower income people is especially the Short-term Rural Cooperative Credit Structure (STCCS) providing mainly short and medium-term credit besides other financial services.

5 At present (March 2005), the three tier STCCS consists, according to statistics of the National Federation of State Cooperative Banks (NAFSCOB), of nearly 1.09 lakh Primary Agricultural Credit Societies (PACS), 368 District Central Cooperative Banks (DCCB) with 12,858 branches and 30 State Cooperative Banks (SCB) with 953 branches or a total of 122,590 service outlets.

On an average, there is one PACS for every 6 villages; these societies have a total membership of more than 120 million rural people making it one of the largest rural financial systems in the world.

PERFORMANCE OF RURAL COOPERATIVE CREDIT INSTITUTIONS

Primary Agricultural Credit Societies

6.04 Primary Agricultural Credit Societies, the credit institutions at the grass-root level deal directly with individual members / clients. A large proportion of PACS also serve as outlets for inputs and for the public distribution system for food and other essential items. The total membership of PACS as on 31 March 2005 aggregated to 1,274 lakh, of which, the borrowing members at 451 lakh constituted around 35per cent (table below).

The total as well as borrowing members of PACS declined during 2004-05. Deposits and borrowings of PACS increased by 5 and 17per cent, respectively, as on 31 March 2005 over the previous year. The loans issued increased by 12per cent during 2004-05 as compared to an increase of 3.3per cent during 2003-04, over the previous year.

Particulars		**PACS**	
(31st March)	**2003**	**2004**	**2005**
No.(lakh)	1.12	1.06	1.09
Members (Cr.)	12.36	13.54	12.74
Borrowers (Cr.)	6.39	5.13	4.51
Owned funds (₹ Cr.)	8198	8397	9197
Deposits (₹ Cr.)	19120	18143	18976
Borrowings (₹ Cr.)	30278	34527	40429
Loan Issued (₹ Cr.)	33996	35119	39212

As per the data available with NAFSCOB, small and marginal farmers constitute nearly 70per cent of total membership of PACS at the national level, while SCs / STs constitute 34per cent.

Wide regional variations are observed – while Western and Southern Regions have a greater proportion of SFs as members, Eastern and Central Regions have nearly half of their members belonging to SC / ST category.

The average membership per PACS (all India) is 1,171 persons as shown in the table below. Southern Region has the highest average at 2,986 persons, while Western has the lowest at 441 persons.

The average borrowing membership per PACS (all-India) is 414 persons. While the Southern Region has the highest average at 944 persons, North-Eastern Region has only 91 borrowing members, on an average.

The gap between the average membership and the average borrowing membership can be presumed as the credit potential available. Viewed thus, there is a huge potential for extending credit outreach by the cooperatives, especially in the Southern Region as shown in the table below.

Though the network of commercial banks and RRBs has spread rapidly and they now have nearly 50,000 branches, their reach in the countryside both in terms of the number of clients and accessibility to the small and marginal farmers and other poorer segments is far less than that of cooperatives.

In terms of number of agricultural credit accounts, the STCCS has 50per cent more accounts than the commercial banks and RRBs put together. Directly or indirectly, it covers nearly half of India's total population.

Sr.No	Region	Total No. Potential	Total PACS (no. 000)	Average Membership /PACS	Total Borr. Membership (no. 000)	Average borr. membership PACS (no./PACS)	Credit membership/ available
1	Central	13273	8322.68	606	6810	496	110
2	Eastern	28928	38952.94	1347	11989	414	933
3	North-Eastern	3628	3835.62	1057	330	91	966
4	Northern	16819	17084.63	1016	7191	428	588
5	Southern	15349	45832.06	2986	14484	944	2042
6	Western	30332	13378.49	441	4266	141	300
	Total	**108779**	**127406.42**	**1171**	**45070**	**414**	**757**

HEALTH OF RURAL CREDIT COOPERATIVES

Despite the phenomenal outreach and volume of operations, the health of a very large proportion of these rural credit cooperatives has deteriorated significantly. The institutions are beset with problems like low resource base, high dependence on external sources of funding, excessive Governmental control, dual control, huge accumulated losses, imbalances, poor business diversification, low recovery, *etc.* Around half of the PACS, a fourth of the intermediate tier, *viz.*, the DCCBs, and under a sixth of the State-level apex institutions, *viz.*, the SCBs are loss-making. The accumulated losses of the system aggregate over ₹ 9,100 crore. Non-performing assets (NPA), as a percentage of loans outstanding at the level of SCBs and DCCBs, at the end of March 2006 were around 16per cent and 20per cent respectively. These institutions do not, therefore, inspire confidence among their existing and potential members, depositors, borrowers and lenders. Thus, there is a need to find ways for strengthening the cooperative movement and making it a well-managed and vibrant medium to serve the credit needs of rural India, especially the small and marginal farmers.

NEED FOR REVIVAL OF THE COOPERATIVE CREDIT INSTITUTIONS

Given the above indicators, there has been considerable debate on whether there is an imperative need for revival of these institutions. Herein, it may be pertinent to reiterate the following points:

In the first place, India is a country with a population of more than 100 crore, of which around 70 crore reside in a little over 6 lakh villages. As there are a little over one lakh PACS in the country, the first fact to be appreciated, therefore, is that every 6th village, on an average, has an existing cooperative credit outlet.

Secondly, although the rural credit system includes about 45,000 rural and semi urban branches of commercial banks and RRBs, if one would net out the cooperatives from the aggregate rural presence of all RFIs, the per outlet population coverage deteriorates from 1:4,393 to 1:14,893. The short point is that on grounds of outreach, cooperatives cannot be ignored.

Thirdly, the outreach is significant not merely in absolute numbers but also in terms of location of outlets. The number of PACS located in hilly terrains, deserts and other areas with poor access far exceed the number of rural branches of commercial banks and RRBs.

The same set of conclusions flow when the system is viewed in the context of its membership. Assuming an artificially low family size of 4, a back of envelope calculation shows that cooperative membership touches the lives of nearly 48 crore rural people which is more than half the aggregate rural population.

As regards number of agricultural credit accounts, the credit cooperatives have 50per cent more accounts than scheduled commercial banks (including RRBs). Of these, 70per cent are estimated to be marginal and sub-marginal farmers.

On the financial side, it is sometimes argued that in terms of deposits, rural and semi-urban branches of commercial banks have nearly six times more deposits than cooperatives. While this is true, it is also true that this is part of a historic policy infirmity which allowed cooperatives to be treated as "refinance windows" instead of incentivising them into becoming genuine thrift and credit institutions.

Coming to the assets side of the balance sheet we find that cooperatives have a 37per cent share in the aggregate crop loans provided by the RFIs. However, in many districts, particularly in remote, hilly and desert areas the share is upwards of 50per cent. 6.18 Notwithstanding a larger client base, the share of cooperatives in institutional credit is lower than that provided by the commercial banks. This is because the average loan size of cooperatives is smaller at ₹ 6,637 as compared to ₹ 31,585 of commercial banks which also confirms the claim that nearly 70per cent of the cooperative structure's clients are small and marginal farmers.

A little known fact is that the system has nearly 64,000 godowns. Given an appropriate policy dispensation entailing private partnership, conditions can be created for holding of crops by farmers at the ground level in these godowns while allowing them to negotiate the godown receipts in the market. A specific plan to revitalize these godowns in cooperative sector be drawn to make them the nerve centres of rural economy on a 'kiosk' model providing storage and financial services.

COOPERATIVES AND THE REFORM PROCESS

In the backdrop of these credentials, a look has to be taken at the reform process and what it means to the cooperative credit institutions. The financial sector reforms have been components of the overall economic reforms undertaken in a phased manner from 1991-92 by GoI. These reforms also envisaged improving the efficiency and productivity of the rural credit delivery system which in turn would accelerate the requisite credit flow to the productive sectors of the economy.

The major objectives of these reforms in the cooperative sector were:

- To make the institutions competitive by removing external constraints having a bearing on their operations,
- To improve their financial health,
- To ensure transparency in their business operations,
- To improve their profitability, *and*
- Institutional building and strengthening.

According to the Task Force on Revival of Rural Cooperative Credit Institutions (Vaidyanathan Committee), the STCCS never realized the enormous potential opened up by its vast outreach owing mainly to a "deep impairment of governance". While they were originally visualized as member-driven, democratic, self-governing, self-reliant institutions, cooperatives have, over the years, constantly looked up to the State for several basic functions.

The Task Force has described in detail how the State Governments have become the dominant shareholders, managers, regulators, supervisors and auditors of the STCCS. The concept of mutuality (with savings and credit functions going together), that provided strength to cooperatives all over the world, has been missing in India. This "borrower-driven" system is beset with conflict of interest and has led to regulatory arbitrage, recurrent losses, deposit erosion, poor portfolio quality and a loss of competitive edge for the cooperatives.

The Task Force Report also recognised that there is an impasse in the laws governing cooperative banking institutions in the country as cooperation is a State subject while banking activities are regulated by a Central Act. Further, the Task Force also took cognizance of the poor quality of internal control systems, house keeping and audit in addition to professionally not qualified human resources manning the CCS. For the revival of the STCCS, the Vaidyanathan Committee Report has suggested an implementable Action Plan with substantial financial assistance for recapitalisation subject to introduction of strict legal and institutional reforms together with technical assistance for human resource development (HRD), establishment of a common accounting system and computerisation.

The implementation of the Revival Package would result in the emergence of strong and robust cooperatives with conducive legal and institutional environment for it to prosper.

At the macro level, the Revival Package is expected to promote growth with social justice and greater financial inclusion. In this background, the following observations are noteworthy:

- With 127 million plus membership and 45 million plus borrowing membership, the STCCS already exceeds the combined client base of commercial banks and RRBs. With nearly 1.25 lakh outlets spread throughout the length and breadth of the country, the STCCS after implementation of the recommendations of the Vaidyanathan Committee would be adequately equipped and would be a highly appropriate mechanism to promote financial inclusion.
- Located in rural areas, governed by local community leadership and managed by the staff with roots in rural areas, the STCCS would be oriented for promoting financial inclusion more effectively.
- The CCS with large membership of small and marginal farmers would have necessary ethos to promote and encourage financial inclusion of deprived sections of the rural economy.
- With limited overheads, the CCS outlets will be an appropriate medium for encouraging financial inclusion through "no frills" savings accounts *etc.*
- The rural populace will be more comfortable and will easily identify culturally with the STCCS.
- Payments under various social welfare schemes of Central and State Governments for instance old age pension, Employment Guarantee Scheme (EGS), scholarships for students, *etc.* could be made more

transparent, safe and foolproof if routed through the STCCS and the recipients could receive the assistance at their doorsteps.

- As envisaged in the Vaidyanathan Committee Report, the membership of PACS should be open to all members of the village without any restrictions. For the purpose, the existing restrictions regarding land holding *etc.*, should be done away with. If need be, PACS could be rechristened as Primary All-purpose Credit Societies instead of Primary Agriculture Credit Societies.
- The STCCS would be an ideal instrument for promoting financial inclusion through SHGs. Many cooperatives could be transformed into SHPIs with proper sensitisation. Promotion of SHGs could, indeed, become a major component of the Business Development Plan of cooperative organisations of the STCCS.
- As the Vaidyanathan Committee is envisaging application of IT right from PACS to SCBs, the STCCS would be in a position to cope with the additional work load on account of promoting financial inclusion. This, in fact, would go a long way in improving the productivity of the credit cooperatives. With due internal controls and checks in place, PACS could ideally be developed as BCs of DCCBs as well as for commercial banks / RRBs wherever the reform process facilitates this.
- As financial inclusion also envisages delivery of financial services at an affordable cost, the STCCS being accessible at the door step of the financially excluded sections of the society will be in a better position to render these services more effectively.
- Located in the rural areas, the STCCS units will have better knowledge of their existing and potential clients and, as such, KYC will not be a major hindrance.

ROLE ENVISAGED FOR COOPERATIVES IN THE PLANNING PROCESS

The cooperative credit institutions are being called upon for a greater involvement in credit dispensation during successive Five Year Plans. The Working Group on Agricultural Credit, set up by the Planning Commission, for the formulation of the X Five Year Plan, had made a projection of credit flow for agriculture at ₹ 736,000 crore at 4per cent growth rate in agriculture GDP. Of this, the cooperative credit structure alone was required to provide during the plan period ₹ 271,000 crore from ₹24,000 crore during the first year (2002-03) of the X Five Year Plan, on the assumption that it would maintain its share of at least 42per cent in the total credit purveyed. During the period 2003-04 to 2006-07, these institutions, put together, have purveyed credit to the tune of ₹ 136,229 crore for agriculture and allied activities.

IMPACT ENVISAGED POST IMPLEMENTATION OF THE RECOMMENDATIONS OF THE TASK FORCE

The Committee concurs with the Vaidyanathan Committee's view that cooperatives are heterogeneous and exhibit wide variations in terms of method of operations, quality of governance, working results, *etc.* The Committee also notes that in certain areas, cooperatives have done exceedingly well.

It is expected that with the implementation of the reform package as envisaged by the Vaidyanathan Committee (already 13 States have signed the MoU with GoI and NABARD, while another 5 States and 1 UT have indicated their "in principle" acceptance), the cooperatives will be able to emerge stronger. Consequent on implementation of the Revival Package, there will be two sets of outcomes – one at the sector level and the other at the macro level. At the sector level the legal environment within which the sector functions will become more enabling and there will be institutional upgradation in terms of systems and procedures, HRD, technology, *etc.* This will improve the internal operational efficiency of the system leading to more robust fundamentals.

At the macro level, this process will complete the financial sector agenda in regard to capitalization of impaired entities. India embarked on its reforms in the financial sector in 1992 with the capitalization of commercial banks followed by that of RRBs. Capitalisation of the rural cooperative credit structure will bring this process to a logical completion.

The other area of impact would be on the national goal of growth with equity. As already indicated, nearly 70per cent of the membership of the rural credit cooperatives are small and marginal farmers. As cooperative institutions improve their health and methods of operation, the number of borrowing members as a percentage of total membership will increase, thus bringing larger number of small and marginal farmers within the cooperative fold. This will help in addressing the credit requirements of this segment of the rural population which is sometimes bypassed in the growth process. This is an important outcome in regard to financial inclusion.

EMERGING ROLE OF COOPERATIVES IN MICROFINANCE

The Andhra Pradesh Mutually Aided Cooperative Societies (APMACS) Act, 1995 has resulted in the creation of over 30,000 new Mutually Aided Cooperative Societies (MACS) in Andhra Pradesh. A vast majority of these are Village Organisations (VOs) of SHGs and about 800 are higher level federations of these VOs at the Mandal level (20-30 villages). All these have been formed as part of the Velugu Programme. However, about 600 MACS are independent of the Velugu Programme, and have been promoted by NGOs. These institutions are able to provide savings and credit services to their members and are fully autonomous.

The parallel Acts are also one way to federate SHGs and form community based MFIs. The founding NGO only plays a facilitating role and there are no external shareholders. Recommendations in regard to the cooperative credit institutions:

EARLY IMPLEMENTATION OF VAIDYANATHAN COMMITTEE REVIVAL PACKAGE

The Committee is in broad agreement with the recommendations of the Vaidyanathan Committee and suggests that all necessary steps should be taken for the early implementation of the STCCS revival package in all States. Consequent on the implementation of the Vaidyanathan Committee recommendations, the Committee hopes for the emergence of a more robust, well managed and self-reliant cooperative credit system with improved governance structures and technology applications.

COOPERATIVES IN SHG LINKAGE – NEED FOR ENABLING LEGISLATIONS

The Committee is of the view that cooperatives are a good forum for enabling financial inclusion through SHGs. This has been demonstrated in several districts such as Bidar in Karnataka, Chandrapur in Maharashtra and Mandsaur in Madhya Pradesh. The Committee notes that in certain States, legislation has been enacted, admitting SHGs as members of PACS and recommends the enactment of similar legislation in other States to enable the emergence of cooperatives as effective SHPIs.

The Committee also recommends that federations of SHGs may be registered in all the States under the Cooperative Societies Act or the parallel Self Reliant Cooperatives Act and availability of funds to these cooperatives for advancing loans may be considered by NABARD, based on objective rating criteria. NABARD may also set aside requisite funds for sensitising the cooperative movement in this regard.

USE OF PACS AND OTHER PRIMARY COOPERATIVES AS BUSINESS CORRESPONDENTS

There are a large number of PACS and primary cooperatives under the parallel Acts located in rural areas where there are no other financial services outlets. Many of these cooperatives are in districts where the DCCBs are defunct or moribund. Such PACS could provide valuable services to their members if they get access to a commercial bank. These PACS could originate credit proposals, disburse loans, collect repayments and even collect savings on behalf of the commercial bank. They could also act as payment channels. RBI has already listed Cooperatives as eligible institutions under the BF/BC Model.

In the circumstances, the Committee recommends that the Cooperatives may make use of this opportunity atleast in States which have accepted the Vaidyanathan Committee recommendations. NABARD may be asked to suggest appropriate guidelines for the purpose, subject to the approval of RBI.

COOPERATIVES ADOPTING GROUP APPROACH FOR FINANCING EXCLUDED GROUPS

Micro-enterprises, in order to be successful, require larger funding which NGOs cannot provide. It will, therefore, be necessary to develop / test a new form of community based organisation other than SHGs which may be more appropriate to support members who engage in micro-enterprises. Those members of SHG who opt to graduate to micro-enterprises could be formed into JLGs or some similar organisation.

Banks may be more inclined to lend to individuals in this group based on the performance of each member in the SHG as well as on the assumption that a JLG will provide some degree of mutual guarantee. There is evidence however, that the relations of mutual trust and support which is described as affinity in a SHG tend to be weaker in a JLG. Therefore, new forms of collateral or guarantee may have to be worked out. In this regard, NABARD has already circulated the guidelines which may be adopted by banks.

Further, the use of the BF model could be thought of to organize vulnerable segments of the population into JLGs. The pilot project presently under implementation by NABARD should be sufficiently broad based to cover the role of facilitators in formation and linkage of JLGs.

RISK MITIGATION - SETTING UP OF CREDIT GUARANTEE FUND

The Committee also recommends the setting up of a Credit Guarantee Fund as a risk mitigation mechanism and also for providing comfort to the banks for lending to such JLGs (akin to the Credit Guarantee Fund Scheme for Small Industries - CGFSI - available for small-scale industries - SSI - at present).

COOPERATIVE CREDIT IN INDIA

Agriculture continues to be main stay of India's national economy. Its contribution to the India's Gross Domestic Product (GDP) is about 30per cent Almost 2/3rd of the population of India depends on this sector for the livelihood. The agricultural land is being taken away by all sectors.

Moreover, in the era of economic liberalization the agriculture is being increasingly recognised as commercial activity, and, therefore, the specific needs of the farmers for modern agricultural technology will, no doubt become complex.

Over a period of years with the state support, the cooperatives have craved out a formidable niche for themselves in various fields of economy like rural credit and banking, fertilizers, sugar, dairy *etc.* There are at present 4.9 lakh Cooperative Societies engaged in diverse business activities with a membership of over 200 million. They have 100per cent villages and 67per cent of rural house holds.

The Cooperative Credit Institutions from the beginning of the 20th Century till 1950s in India have depended heavily for their credit needs on individual sources of finances such as moneylenders, traders, commission agents, landlords, relatives and friends. There sources by their very nature, were not capable of meeting the growing credit needs of the farmers. Moreover, the farmers were subjected to many unfair conditions including higher interest rates. Realizing these defects, both the Central and the State Governments in India took a number of policy decision and enacted necessary legislation to progressively institutionalize farm finance. Today, institutional credits accounts for a share of over 50per cent of total credit need compared to less than 10per cent in the fifties

The Flow of credit to the agriculture sector and the rural areas is met through institutional agencies like Commercial Banks, Regional Rural Banks and Cooperative Credit Institutions. The short term, medium term and long term needs of the farming and rural community are met by their credit institutions. The Cooperative Banks have played a dominant role in providing credit to the needy.

SHORT TERM AND MEDIUM TERM FINANCE

Strengthening financial resources base of the various cooperative credit institutions and enterprises is definitely a priority area for working out a competitive edge for cooperatives in the field of short term and medium term credit system. The cooperative banks like PACS at the fields of short term an medium term credit system. The cooperative banks like PACS at the primary level have to show a turnaround through implementation of Business Development Plans (BDP).

At the same time it is necessary to find out how far the implementation of BDPs have helped in shoring up the finances of the cooperative banking system through reduction of loan overdues? How far the PACs associated with losses have reappealed into profits in the wake of the implementation of various components of BDPs?

LONG -TERM FINANCE

In the field of long term finance the state ARDBs are providing long term investment finance to the farming community in the rural areas for financing minor irrigation, farm mechanization, land development projects, fisheries, poultry *etc.* The state ARDBs by and large are dependent on re-finance facilities extended by NABARD. The State ARDBs have to make serious efforts to increase their financial health through mobilization of deposits and in the process reduce their dependence on re-finance facilities of NABARD. In this context, they have to evolve attractive deposit schemes and at the same time, will have to evolve Deposit Insurance Schemes (DIS) to instill confidence among depositors.

APPLICATIONS OF PRUDENTIAL NORMS:

The RBI has directed the State Cooperative Banks to adopt prudential norms in terms of income recognition, assets classification, and provisioning for bad and doubtful debts. Prudential norms also includes capital adequacy norms and these norms have been fixed at 8per cent of the capital risk weighted assets. The Cooperative banking sector has been asked by RBI to adopt prudential norms by March 31st, 2000. Similarly, the Urban Cooperative Banks which constitute a self-reliant group of banks in urban area have been asked to adopt prudential norms. Adoption of prudential norms by Cooperative Banking sector would imply considerable improvement in their financial health, reduction of loan overdues and greater mobilization of deposits in the coming years. The Non-performance Assets (NPAs) associated with cooperative banks have to be drastically reduced through appropriate measures. This process would help in the recycling of funds for lending to the farming community. Future lending should be based on careful analysis of loan application including risk assessment to be backed by ensuring utilisation of funds for the specified objectives.

GOVERNING CORPORATE IN COOPERATIVES

In the context of intensive competitive economic environment introduced in the wake of economic reforms, the Governing Corporate system in cooperatives is likely to undergo change in the coming decades of the 21st century. The national Cooperative policy which is under formulation in the Urban Ministry of Agriculture is likely to lay emphasis on gradual disengagement of Government from financing of various cooperative activities and cooperative enterprises. Keeping this scenario in view, the cooperative have to adopt stringent measures to improve their financial position through greater deposit mobilization and access to new ways of raising capital from capital market. In this context they also have to rely on cost reduction and cost effective technologies to control transaction costs as well as management costs. Cooperative Leaders have to work out on action plan for removing various restrictions which have been have been placed by the Government for the smooth functioning of cooperative enterprises. In this Action Plan, greater emphasis would have to be laid on members participation in decision making process and in arranging restructured Education and Training Programmes for the members and personnel engaged in cooperative development.

Governance by the cooperative leaders would imply greater emphasis on promotion of information technology in various segments of cooperative movement.

COOPERATIVE MANAGEMENT

Access to latest information and policy decision having a bearing on the future of Cooperative movement will have to be constantly tied up with the information matrix availability to the cooperative management system. To safeguard their interests, the cooperative leaders will have to establish lobby with the help of opinion on leaders and members of parliament to safeguard the interest of cooperatives in the coming years.

For improving the efficiency parameters of cooperative management there is the need to remove false members, false leaders and false cooperative enterprises from the domain of cooperative Movement. Such a step is necessary to lift the cooperative movement out of the morass into which it has fallen in certain region of India.

The efficient management of cooperative would also imply the implementation of the policy relating to mergers and acquisitionary among cooperative enterprises. Attention should be given to information Technology at primary level to improve the Cooperative Management Information Systems. In the future years great stress would have to be laid on encouraging qualified dedicated members to fight the election for the top leaders of the Board of Management and Governing Council of Cooperative. This would help in improving the cooperative face the competitive economic environments with courage and determination.

REASONS FOR DETERIORATION IN THE COOPERATIVE BANKING SYSTEM

Short-term and Medium Credit Finance

- The base structure of agriculture credit -cooperative consisting of PACS at the ground level continues to be weak even after recognizing and revitalization schemes including BDP being implemented since, 1991. The impact of the implementation of BDP on the functioning of PACS needs consideration.
- The level of overdues in the Cooperative Banks continues to be an area of concern.
- The central government should extend adequate budgetary support to strength the capital base of the weak and non-viable cooperative banks in the near future.
- Various disciplines pertaining to credit limits for seasonal agricultural credit limits are imposed by NADARD. There conditions act as constraints are full utilisation of credit limits sanctioned by NADARD. Relaxation of there constraint needs to be considered.
- The Cooperative Banks should take advantage of the deregulated interest rates on deposits and lendings. The State Government should

not interfere in the fixation of the interest rates by the Cooperative Banks .

- The PACS should diversify their operations into non-credit activities like jewel loans, consumption loan *etc.* Keeping in the view requirements of the rural masses in their area of operation.

Long -Term Credit/ Finance

- The resource base of ARDBs needs to increased through mobilization of deposits and rural savings. This would require strengthening of owned funds by ARDBs. Also, further avenues for resource mobilization by ARDBs may have to be identified and discussed.
- Remedial measures for improving the recovery position may be considered and discussed.
- NABARD imposes rigid discipline on ARDBs while extending re-finance support. Their mobilization needs to be mobilized.
- Prudential norms have been made applicable to ARDBs with effect form accounting year 1997-98. The ARDBs should be try to classify their assets into specific categories like standard, sub-standard, doubtful and loss making assets. Thereafter adequate provisioning has to be made to take care of the risk elements involved in various types assets.
- Training of staff in non-farm finance and housing finance may be considered and discussed. The cooperative leaders and members need to be acquainted with the techniques such as project appraisal and project formulation. This would be in the interest of proper appraisal of loan applications by the ARDBs and would help in the reduction of loan overdues.

The State partnership, by way of government contribution to share capital, which was conceived as a measure for strengthening the cooperative credit institutions, has passed the way instead for an unnecessarily high level of state control, their increasing bureaucratization and politicization, culmination in virtually depriving the cooperatives of their inherent vitality and affecting their democratic and autonomous character. In the milleu of liberalization and privatization, there is an urgent need to redefine the Government's role in developing and strengthening cooperative banks. The various reforms initiated in different sectors, are yet to be introduced to the cooperatives.

MEASURES TAKEN TO FACE THE EMERGING CHALLENGES

Cooperatives are based on the principles of autonomy and independence and strive for becoming member owned and member controlled organisations. In putting this principles in practice, NCUI has been making concerted efforts to persuade the Government of India and the State Governments to remove the restrictive provision from the existing Cooperative laws and to modify them in tune with Model Cooperative Act enunciated in the Ch. Brahm Prakash Committee Report. There are reports that the central Cabinet has approved amendments in the Multi-State Cooperative Societies act 1984. However, these amendments are yet to be approved by the parliament. Similarly some State Governments like Andhra Pradesh have passed mutually Aided Cooperative Societies Act under which there is no interference from the government in the functioning of the cooperative societies. Other State governments like M.P., Rajasthan, Punjab are reported to have initiated steps to modify State Cooperative acts in tune with the provision of Model Cooperative Law. With the approval of the Government of India and RBI, the Cooperative bank of India has been established to bridge the systemic gaps existing within the cooperative credit structure . However due to lack of an amendment in the Banking Regulation Act, it has not been possible so far to operationalise the functioning of the Cooperative Bank of India. Efforts are being made to persuade the Government of India and the Ministry of finance to issue a license to this bank so as to enable it to operationalise its functioning.

The government has extended a sizeable budgetary support to the public sector commercial banks for strengthening their capital base and for enabling them to implement prudential norms. Similar financial assistance to the weak and non-viable cooperative banks has not been extended by the government of India.The cooperative leaders feel that in the background of a level playing field approach, the Cooperatives should also be extended a reasonable level of budgetary support so as to enable them to clear their balance sheets. Such a process would help the weak cooperative banks to strengthen their financial resource base and would enable them to stand on their own feet to face the competitive challenges in the liberal market environment.

Emphasis is already being laid on building up strong human resource base in the cooperative sector. Efforts are being to lay greater emphasis to promote professionalisation in the management and training of the cooperative personnel so as to enable them to man the various post in the cooperative organisation. The training and the evaluation programmes are being restructured on scientific lines after capturing the core elements of the information technology and management information system. It is being recognised that only highly professionalized management cadre and trained cooperative personnel would be able to face the competitive challenges from the private sector. Keeping this aspect in view the various training programmes are being extended by VAMNICOM and other ICMS with a focus on professionalisation in management and efficiency of cooperative personnel.

RURAL CREDIT COOPERATIVES IN INDIA

Rural Credit Cooperatives have existed in India for a long time. A shortage of supply of rural credit was prevalent in India. To meet the demand for short and long term rural credit the Co-operative Credit Structure (CCS) was set up.

While short term credit is supplied by the State Cooperative Banks (SCB), District Central Cooperative Banks (DCCB) and Primary Agricultural Credit Societies (PACS), long term credit is supplied by the Primary Cooperative Agriculture and Rural Development Banks (PCARDB).

Rural Credit Cooperatives were initiated in India long back, some of them, even before India's Independence in 1947. Post-independence the rural credit cooperative system was developed further. Moreover rural banks were set up. However in spite of such initiatives credit needs of the rural Indian people have not been met effectively. Supply of credit for agriculture too has not matched up to the demand levels.

The resultant effect has been widespread discontent and despair among the rural poor, sometimes leading to extreme actions like suicides by farmers.

Measures were undertaken by the Indian government in 2004, to increase credit supply for agriculture by commercial banks. Moreover banks were also asked to re-organise repayment schedules of farmers who were affected by floods, droughts, *etc.* According to World Bank estimates of 1994 and 1995, the average credit utilisation was ₹ 14,549 per family in a year. While 65per cent of rural credit in India was utilized for productive purposes, the remaining 35 per cent was utilised for consumption purposes.

Out of the 65per cent utilisation for productive purposes, short term utilisation constituted 49per cent. Long term utilisation (purchase of agricultural machinery, livestock) etc accounted for a mere 16per cent. Short term consumption (for purchase of consumer durables, clothes, etc) was 20per cent while long term consumption (for house building, marriage, *etc.*) was 15per cent.

The Co-operative Credit Structure (CCS) of India was set up to serve the needs of both short term and long term rural credit in India. Short term credit is supplied in rural India by three institutions -

- State Cooperative Banks (SCB)
- District Central Cooperative Banks (DCCB)
- Primary Agricultural Credit Societies (PACS)

Long term credit is supplied by the Primary Cooperative Agriculture and Rural Development Banks (PCARDB).

In 2007, the World Bank has approved a project related to rural credit cooperatives in India. This project entitled "The Strengthening Rural Credit Cooperatives Project" involves a total cost of an estimated 600 million US dollars and is expected to be completed by June 2012. The National Bank for Agriculture and Rural Development (NABARD) will be implementing the project.

CREDIT CO-OPERATIVES ARE TOO BIG TO WORK

It has become commonplace to hail institutions with a hoary past, irrespective of their fundamental weaknesses. However, such superfluous praise can do more harm than good to the cause of the institution and its stakeholders.

A case in point was when a top functionary of the RBI recently glorified the rural credit co-operatives for following the tenets and principles of co-operation. However, it is the failure to adopt the principles and tenets of co-operation that is the bane of credit co-operatives in India.

The cooperative should be an effort driven by members and there should be joint liability to avoid free rider problems. Members have been unable to enforce credit discipline because of the extremely large size of these institutions. This, in fact, goes to explain the success of the self-help group model. Credit co-operatives were, in fact, originally conceptualised as a small group where, more than material means, the moral pressure which the members exert on each other plays a key role in their success. Peer pressure ensures recovery and recycling of funds. It is, however, diluted as membership size grows.

The Mac Lagan Committee, in 1914, had suggested that optimal group size be limited to 40 members to ensure cohesion and peer pressure. This basic principle of cooperation was, however, turned upside down in the post-Independence era in India. The average membership size of primary agriculture credit societies is around 1400.

STATE CONTROL

In post-independent India, rural credit co-operatives evolved more as state sponsored initiatives than member-driven ones. The implementation of the recommendations of the All India Rural Credit Survey (AIRCS) Committee, which submitted its report in 1954, shaped the future of cooperatives in India. The Committee recommended state partnership at all levels of co-operatives, in the form of share capital contribution, provision of technical, managerial and financial assistance to co-operatives. Cooperatives came to be perceived as a state agency, rather than an autonomous, member-based economic enterprise.

The Mirdha Committee, while assessing the growth of the co-operative movement in 1965, had observed that the government policy of deliberate expansion of cooperatives had led to their politicisation and the entrenchment of vested interests. The Committee observed that cooperatives had drifted far away from their original objectives. Based on the observations of the Mirdha Committee, the Conference of State Ministers of Co-operation, in 1969, recommended stringent provisions in cooperative legislation like to curb vested interests. The restrictive provisions, in effect, legalised government interference, which reached its nadir in 1977 when democratically elected management committees of cooperatives were superseded in nine States with a change in government at the Centre.

VAIDYANATHAN PANEL

The 1990s witnessed attempts to unshackle the cooperative sector from the strictures of the government and restore their democratic character. First, was the attempt by an Expert Committee under the Planning Commission of India which submitted a Model Cooperative Bill. Progress in implementing the Model Law was tardy because of the States' reluctance to shed their control. Subsequent committees on cooperatives, notably those headed by Jagdish Capoor (1999), Vikhe Patil (2001) and V.S. Vyas (2002) strongly supported replacing existing laws with the proposed Model Law.

The latest attempt in the bid to refurbish the credit cooperative system, was the Vaidyanathan Committee in 2005. The Vaidyanathan Committee in line with its mandate suggested an Implementable Action Plan with substantial financial assistance for recapitalisation. Recapitalisation assistance, however, was made conditional to the introduction of strict legal and institutional reforms, capacity building, establishment of a common accounting system and computerisation. In order to avail assistance under the revival package, the State government concerned was required to enter into a memorandum of understanding (MoU) with the NABARD and the central government. Till date 25 States have entered into MoUs. However, there has been a certain amount of dilution in the stipulations set by the Vaidyanathan Committee in the MOUs.

For instance, the Committee had recommended the retirement of government equity over time, but subsequently it has been decided to retain 25 per cent of government equity in cooperatives. More importantly, the eligibility conditions to avail financial assistance under the revival package have also been diluted.

It has been more than six years and the recommendations of the Committee are yet to be fully implemented. Further, during the implementation of the revival package, agricultural debt waiver scheme of the Government announced in 2008 has created serious moral hazard problems. With dilution of the original recommendations, tardy implementation and debt waiver scheme of the government, it is no surprise 43 per cent of the 94,647 village level Primary Agricultural Credit Societies (PACS) in the country were loss-making as on March 2011 and that overdues of cooperatives in 2010 were much higher at 41 per cent of loans given, compared with 34 per cent in 2005.

However, the Vaidyanathan Committee has not addressed the issue of ideal member size of co-operatives. Without according due respect to the basic principles governing them, it is unrealistic to expect anything better from the credit co-operatives.

PRODUCTION CO-OPERATIVES

The co-operative societies which deal with the activities related to agricultural and industrial production are known as Production Co-operatives *i.e.* Farming, Industrial and Processing Co-operatives which are found on large scale in all over India.

FARMING CO-OPERATIVES

"Co-operative Farming necessarily implies pooling of land and joint cultivation and management".

The term Co-operative Farming is often used as a farm management in which land is jointly cultivated. In other words, application of the principles of co-operation in the cultivation of land is called co-operative farming.

For a country like India, the problem of agricultural production is of utmost important, where land is scarce and even now concentrated in a few hands, yields are low but rents are high, farmers are poor and farms expensive, a thorough overhauling of the tenure system is called for.

Thus, a co-operative farming society is a voluntary organisation of farmers based on ideals of self-help and mutual aid. With the objectives of increase in production and employment, improvement in techniques of production, rationale use of land and most desirable allocation of man-power resources.

According to the Planning Commission (1961), the main features of the co-operative farming- Members pool their lands, manpower and other resources willingly in a single unit. Since, it is a voluntary association, members can withdraw any time. The farming is done on co-operative lines *i.e.* it is cultivated jointly. Ownership of the land generally lies in the hands of the individual members *i.e.* the right to land is never surrendered. Management is jointly conducted by a committee elected from the members. The programme of farm operations is laid down by the society and each member does his assigned work according to the scheme. Members receive share in the produce according to the work put in and the land contributed for joint cultivation. The object is to promote efficient production, purchase and sale of farm produce and supplies as well as to meet other needs of farmers. And net profits are utilised for payment of bonus to members after providing reserves.

Table. Progress of Farming Co-operative Societies in India.*

(Membership in 000′ & Value in Million)

Year/ Items	No. of Societies	Membership	Share Capital	Working Capital	Turnover
1994-95	6652	328.4	67.4	502.5	182.3
1995-96	6915 (3.95)	316.5 (-3.62)	71.6 (6.23)	602.1 (19.82)	264.7 (45.05)
1996-97	7387 (6.82)	359.2 (13.49)	77.2 (7.82)	622.1 (3.32)	293.6 (10.98)
1997-98	7199 (-2.54)	313.6 (12.80)	74.6 (-3.37)	444.1 (-28.61)	(20.82) 232.2
1998-99	7349 (2.08)	344.5 (9.85)	83.8 (12.33)	488.5 (10.00)	404.7 (74.14)
1999-00	7133	352.2	78.8	494.7	(23.02)

	(-2.94)	(2.23)	(-5.96)	(1.27)	311.9
2000-01	7001 (-1.85)	350.1 (-0.59)	76.5 (-2.92)	496.9 (0.44)	312.2 (0.32)
2001-02	7146 (2.07)	344.1 (-1.71)	67.6 (-11.63)	535.7 (7.81)	322.2 (3.20)
2002-03	7113 (-0.46)	343.7 (-0.12)	70.9 (4.88)	628.2 (17.27)	329.3 (2.17)
2003-04	7116 (0.04)	343.6 (-0.03)	70.9 (0.00)	628.3 (0.02)	329.1 (0.00)

*Indian Co-operative Movement A Profile- 2006, National Resource Centre, National Co-operative Union of India, 2006. P.47.

During the Five Year Plans emphasized the importance of the multi-purpose society for the rural area and the need for each village to have a co-operative organisation to cater for the multiple needs of the members. Every encouragement was given for the establishment of co-operative farms. Particularly, small and medium farmers were to be encouraged and assisted to group themselves voluntarily in co-operative farming societies. By 1955-56, there were about 1000 co-operative farming societies with a membership of 1.9 lacks in India, mostly in Panjab, Bombay and Utter Pradesh. A sum of ₹ 50 lack was provided for promoting this societies. By 1960-61 the number of co-operative farming societies increased to 6,325 and membership to 3.05 lacks and the working capital to Rs.6.90 lack and the land under cultivation was 8.90 lacks acres.

In 1965-66, the number of farming co-operative was 7,294 with a membership of 1.88 lacks and a cultivated area of 3.9 lack hectares. In June 1971, there were 9,473 farming co-operatives with a membership of 2.57 lacks and area covered 4.88 lack hectares. Out of these 5,070 were joint farming co-operative and 4,403 collective farming societies with a membership of 1.2 lacks and 1.35 lacks with covering an area of 2.87 and 2.01 lack hectares, respectively.

The table above shows the progress of farming co-operative societies in India for last ten years *i.e.* during the period of economic reforms. Number of societies increased from 6652 in 1994-95 to 7116 in 2003-04. Therefore, growth rate of these societies was uneven during the same period as it was 3.95 at 1995-96 decreased up to 0.04 in 2003-04. It was negative during the years 1997-98, 99-00, 2000-01 and in 2002-03. Membership increased from 328.4 to 343.6 during for the period and share capital increased from ₹ 67.4 million in 1994-95 to ₹ 70.9 million in 2003-04. Table shows that the membership is also not even. Working capital increased from ₹ 502 million in 1994-95 to ₹ 628.3 million in 2003-04. As well as, turnover of the farming co-operative societies was increased from ₹ 182.3 million in 1994-95 to ₹ 329.1 million in 2003-04, shows around two times increase in total turnover. Therefore, the growth rate of share capital, working capital and turnover is shown a declining trend along with negative rate.

Farming societies were mainly based on voluntary co-operative principle with two objectives of realizing economic units and securing the development of the rural economy, for assistance to the poorer sections.

Reasons of limited success of Co-operative Farming in India are such as-lack of educated and enlightened leadership in the rural areas, lack of social consciousness among the members, too much emphasis on membership rather than on quality, failure to create confidence and enthusiasm among the workers, preponderance of absentee land owners as members, lack of co-operative spirit in the village life, delay in realizing State funds to the societies.

INDUSTRIAL CO-OPERATIVES

An Industrial co-operative consists of a co-operative formed by artisans, craftsman, industrial laborers and small industrialists either of undertaking production and marketing or for providing facilities and services to them. Broadly, industrial co-operatives can be divided into two types; i) societies that undertake production and ii) societies which provide service to their members. Production societies are mainly primary societies while service society's functions are undertaken both by the primary and the federal societies.

Industrial co-operative societies have set up their federal bodies. These federal bodies are organised on single industry as well as on multi-industry basis.

Industrial federations exist at the District and State level. State level federations admit primary co-operative societies as their members. District and State level federations have federated into national level bodies. Federations render several services *e.g.* supply of basic raw materials, finishing materials, evolving new patterns of designs, sale of finishing products, *etc.* In India industrial Co-operatives get financial assistance from the Government, Apex and Central Co-operative Banks, Industrial Co-operative Banks, State Bank of India, Reserve Bank of India, and Commercial Banks. The Government provides financial assistance through loans and also participates in the share capital.

Ministry of Industrial Development has classified the societies such as-Weavers Co-operatives, Spinning Mills, Industrial Co-operative Banks, Industrial Estate, and Others which include Paddy, Pottery, Oil Crushing, Canning of fruits and vegetables, Chemical Industries, Handicrafts Industries, Leather Goods, Construction Materials, Sericulture, Spinning Societies, Coir and Miscellaneous Industries. After Independence, the Industrial Policy Resolution of 1948 gave special attention on cottage and small scale industries.

The Resolution envisaged that in the mixed economic order, where the key industries will be in the public sector, the village and cottage industries would be organised and developed on co-operative lines. Afterwards the

Industrial Policy Resolution of 1956 emphasized the formation and development of industrial co-operatives in the country.

As a result of various measures such as setting up of separate Boards for different industries, liberal financial assistance in marketing of products and programmes of improving the productivity of the workers *etc.* the organisation of industrial co-operative societies received a definite encouragement and from 1951 to 1971-72, as a result the number of societies, their membership, working capital and sales *etc.* increased considerably. Following table shows the progress of industrial co-operatives in India.

Table. Progress of Industrial Co-operatives in India.*

(₹ in Crore)

Year	No of Societies	Membership (000)	Working Capital	Sale
1950-51	7101	766	7.05	N.A.
1960-61	33266 (368.5)	2564 (234.7)	44.42 (530.0)	69.85 (–)
1970-71	46640 (40.2)	3022 (17.8)	212.93 (379.3)	163.00 (183.3)
1971-72	48000 (2.9)	3300 (9.2)	220.00 (3.3)	173.00 (6.1)

*Memoria C.B., Kamat G.S., "Co-operation: Theory and Practice", Kitab Mahal, Allahabad, 1975. P.68.

The data in above Table above substantially brings out thc progress of industrial co-operatives during the initial two decades of the planned India. It increased from 7101 in 1950-51 to 48000 in 1971-72 whereas the membership increased from 766 thousand to 3300 thousand and working capital ₹ 7.05 crore to ₹ 220.00 crore; growth rate of the societies was uneven and compare to year 1960-61 it is shown declining trend, therefore, it was positive. Sale of industrial co-operatives increased from ₹ 69.85 crore to ₹ 173.00 crore during 1960-61 to 1971-72. Therefore, number of societies, membership, working capital and sale of industrial co-operatives is shown a declining trend under these twenty years span period.

Table above shows the progress of Weavers and Non-Weavers Co-operatives in India. Number of weavers co-operatives increased from 20,940 to 20,871 and non-weavers co-operatives from 49,142 to 48,841 during 2002-03 to 2003-04 with growth rate of -0.33 and -0.61 respectively. For the same period the membership increased from 19177 to 19191 thousand and 2255100 to 2256543 thousand whereas growth rate was 0.07 and 0.06 respectively. Share capital of weavers increased from ₹ 1611.38 to ₹ 1608.78 million and non-weavers increased from ₹ 951 to ₹ 964 million during the period from 2002-03 to 2003-04.

Table. Progress of Weavers and Non-Weavers Co-operatives in India.*

(Value ₹ in Million)

	Weavers Co-operatives		Non-Weavers Co-operatives	
Items	**2002-03**	**2003-04**	**2002-03**	**2003-04**
No. of Co-operatives	20,940	20,871 (–0.33)	49,142	48,841 (–0.61)
Membership (000')	1,91771	1,9191	2,255100 (0.07)	2,256543 (0.06)
Share Capital	611.38	1608.78 (Negligible)	951.1	964.7 (1.43)
Govt. participation in share Capital	36.6%	40.03% (3.43%)	34.7%	34.2% (–0.5)
Working Capital	9113.9	9041.1 (Negligible)	4400.5	4410.5 (0.02)
Total assets	5130.7	5402.2 (5.29)	1495.0	1486.7 (–0.56)
Value of product sold	14,687.8	14,458.3 (–1.56)	5297.0	5301.5 (0.08)

*India co-operative movement a profile- 2006, National Resource Centre, National co-operative Union of India, 2006. P.44.

Government contribution in share capital of weavers co-operatives was 36.6per cent in 2002-03 which increased to 40.03per cent in 2003-04 and for non-weavers the same increased from 34.7per cent in 2002-03 to 34.2per cent in 2003-04.

Working capital of weavers co-operatives decreased from ₹ 9113 to ₹ 9041.1 million and non-weavers working capital increased from ₹ 4400.5 to ₹ 4410.5 million during the same period. Growth rate of total assets and value of product sold was also very less. Therefore, table shows that the progress is uneven and not much more remarkable.

Data of the Table above shows the production of fabrics or cloth in mills of the category of Handlooms and Power looms during the last decade. Production of mill sector was 1957 million square meter (6per cent to total) in 1996-97 which decreased to 1493 (3per cent to total) in 2005-06. Handloom production was 7403 million square meters (19per cent) in 1996-97 but decreased to 6087 million square meters (13per cent) in 2005-06. For the same period Power Loom production increased from 19352 million square meters (56per cent) to 29627 million square meters (62per cent).

Total production of fabrics cloth in mills *i.e.* handlooms and power looms increased from 34813 million square meters to 47730 million square meter for the above stated period. As well as in total production of fabrics power loom production was highest compare to mill sector and hand loom. However, Industrial Co-operatives suffered of several problems; such as- inadequate supervision and inspection and lack of prompt audit, irregular and insufficient supply of raw materials, fluctuations prices, insufficient storage facilities, lack of cheap and adequate finance, accumulation of unsold stock, monopoly of middlemen and traders, *etc.* The over-all picture of working of industrial co-operatives for the reforms period has been discouraging.

Table. Production of Fabrics/Cloth in Mills, Handlooms & Power looms.

(Million Square Meters)

Years/Item	Mill Sector	Hand Loom	Power Loom	Total
1996-97	1957 (6%)	7403 (19%)	19352 (56%	34813
1997-98	1948 (5%)	7603 (19%)	20303 (55%)	37441
1998-99	1785 (5%)	6792 (19%)	20690 (57%)	36102
1999-00	1714 (4%)	7352 (19%)	23187 (59%)	39208
2001-02	1546 (4%)	7585 (14%)	25192 (62%)	42034
2003-04	1434 (3%)	5493 (13%)	26947 (64%)	42383
2004-05	1503 (3%)	5722 (13%)	28325 (62%)	45355
2005-06	1493 (3%)	6087 (13%)	29627 (62%)	47730

Note:

Figures in parentheses show the percentage to last coloum (total).

PROCESSING CO-OPERATIVES

Processing co-operatives in India began working with the setting up of a ginning unit in 1917. Afterwards Sugar Co-operatives, paddy husking, groundnut decorticating got also added to the sector. Co-operative processing units are registered under the Co-operative Societies Act of the concern State and up till now work is regulated by co-operative rules and by-laws. The membership consists of individual agricultural producers, co-operative marketing societies and Government service co-operatives located in the area. In order to enable processing units require block capital for construction of building and installation of plant and machinery which is raised through contribution by the members to share capital, State contribution to the share capital, medium/ long term loans from the Industrial Finance Corporation, State Co-operative Banks and State Bank of India.

More co-operatives in the field of jute, oil crushing, fruit and vegetables, copra, teas were set up during the plan periods. Processing co-operatives gained special attention during the plan periods. Third Five Year Plan, considered the development of co-operative processing as an essential not only for increasing rural incomes and facilitating credit for production, but also for building up a through co-operatives rural economy.

Table. Progress of Sugar Co-operatives in India.

(Value ₹ in million)

Items/Years	2003-04	2004-05	2005-06
No. of Installed sugar factories	316	315 (–0.32)	316 (–0.32)
No. of factories in operation	235 (–13.62)	203 (17.73)	239
Membership(million)	5.264 (–2.43)	5.136 (0.00)	5.136
Share Capital	N.A. (–)	33,235.7 (–3.03)	32,226.4
Working Capital	N.A. (–)	87.366.1 (0.00)	87,366.1
Turnover(Total)	N.A. (–)	12,808.0 (0.00)	12,808.0

*Indian Co-operative Movement A Profile- 2006, National Resource Centre, National Co-operative Union of India, 2006. P.37.

The above Table above brings out the performance of Sugar Co-operatives in India between the periods of 2003-04 to 2005-06. In all the progress of co-operative sugar factories for the reforms period does not bring out any significant progress.

The number of installed factories remained stagnant whereas for membership shows little decline; share capital of them has decline whereas the working capital and turnover remained constant. This is not only because of the economic reforms but overall economic problems that have been faced by the Indian agriculture. Sugar co-operatives in India in general and that of Maharashtra in particularly served as centers of the several economic growths. The role played by sugar co-operatives has remained significant even during the period of economic reforms.

The above Table above shows the share of co-operatives in total sugar production and capacity of utilisation. Production by co-operatives during the period of 1995-96 to 2004-05 shows fluctuating trend. Compared to production of co-operative sugar factories during 1995-96 were 9.6 million tones decreased up to 4.7 million tones during 2004-05.

Table. Co-operatives Share in Total Sugar Production.*

(Sugar Production Million in Tonnes)

Items/Years	Total Sugar Production	Production by Co-operatives	Co-operative share to total production in %
1995-96	16.5	9.6	58.6%
1996-97	12.9 (–26.03)	7.0 (–22.08)	54.1% (–7.68)
1997-98	12.9 (0.00)	7.1 (1.43)	54.9% (1.48%)
1998-99	15.5 (20.15)	9.0 (26.76)	57.7% (5.10)
1999-00	18.2 (17.42)	10.4 (15.55)	56.9% (–1.38%)
2000-01	18.5 (1.64)	10.5 (0.96)	56.7% (–0.35%)
2001-02	18.5 (1.64)	9.4 (0.96)	50.8% (–0.35)
2002-03	20.1 (8.64)	10.2 (8.51)	50.5% (–0.59)
2003-04	13.5 (–32.83)	6.0 (–21.66)	44.4% (–12.08%)
2004-05	12.7 (–5.92)	4.7 (–21.66)	36.7% (–17.34)

*Indian Co-operative Movement A Profile- 2006, National Resource Centre, National Co-operative Union of India, 2006. P.37.

Therefore, the total sugar production in India which is including private, public and co-operative sector production is also showed uneven trend during the same period. Co-operatives share in total production which shows decline trend as it was 58.6 million tones in 1995-96 decrease up to 36.7 million tones in 2004-05.

Only excluding 1997-98 and 1998-99 during other year's co-operative share to total sugar production was negative during the 1995-96 to 2004-05. Table revealed that the production of co-operative sector and share of co-operatives to total production is decreased during the same period.

RURAL COOPERATIVE BANKS IN EXTENSION OF AGRICULTURE CREDIT

Institutional lending to agriculture and allied activities has recorded a significant growth during the last two decades due to the aggressive policy of the government in expanding credit to agriculture. The following figure depicts the growth in agricultural credit distributed by the STCCS:

Table. Performance of STCCS under Agricultural Credit.

(₹ in crore)

Years	StCBs	DCCBs	PACS
2002-03	35052	52820	33996
2007-08	53314	95974	57642
2008-09	51866	97207	58787
2009-10	59784	104997	74938
2010-11	68481	122795	91304
2011-12	81523	144761	107300

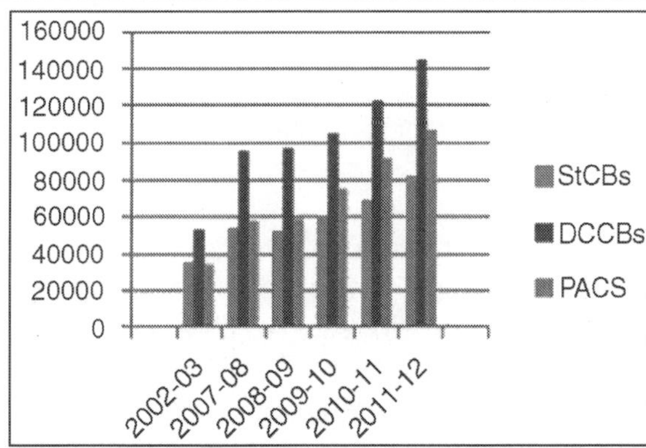

Table. Agricultural Loans Disbursed during the Year.

(₹ in crore)

Agency	2006-07	2007-08	2008-09	2009-10	2010-11	2011-12
Coops.	42,480 (18)	48,258 (19)	46,192 (15)	63,497 (17)	78,121 (17)	87,963 (17)
RRBs	20,435 (9)	25,312 (10)	26,765 (9)	35,217 (9)	44,293 (9)	54,450 (11)
CBs	1,66,485 (73)	1,81,088 (71)	2,28,951 (76)	2,85,800 (74)	3,45,877 (74)	3,68,616 (72)
Total	**2,29,400**	**2,54,658**	**3,01,908**	**3,84,514**	**4,68,291**	**5,11,029**

Lending by cooperative credit institutions grew much faster than that by other banks during the 1990s. Thereafter, their share has been declining since, 2007-08.

Despite this fall, cooperatives clearly remain the major source of agricultural credit to small and medium farmers in terms of the number of borrowers and size of individual loans.

Figures in brackets indicate percentage share of different agencies to total agricultural credit

PROBLEMS OF COOPERATIVES

Despite the phenomenal outreach and volume of operations, the health of a very large proportion of rural credit cooperatives has deteriorated significantly. The institutions are beset with problems like poor governance infrastructural weaknesses, operational inefficiencies and the consequent impairment of their financial health. Several factors such as low borrowing membership, low resource base, lack of democratisation and professionalism, high incidence of overdues and almost stagnant recovery performance have led to the deterioration in the financial soundness of cooperatives. There is an urgent need to find ways for strengthening the cooperative movement to meet the credit needs of rural India, especially the resource-poor and resource-less poor farmers. The revitalisation and strengthening of cooperative institutions at all levels should therefore be considered not only desirable but expedient. The thrust has to be four-fold, financial, operational, organizational and systemic.

POOR GOVERNANCE

Cooperatives operate within the legal framework formulated by state governments. However, compliance with the legal provisions of the State Cooperative Societies Acts has not been ensured. Governance, connected lending, transparency in grant of loans, audit, internal checks and control, recovery of dues, recruitment of qualified persons are issues trammeling the efficient functioning of cooperatives. As dominant share holders, state governments interference in the management of cooperatives. Supervision and guidance by the elected Boards is lax. Delay in conduct of elections, frequent supersession of the Boards, lack of participation by members in the management and decision-making process have impaired functioning of cooperatives on sound business lines. In the absence of professional management, accountability and uncertain tenures, the Board members are not able to provide dynamic leadership to the organisation. Restoring and strengthening autonomy, mutual help and self governance are the cornerstone of the cooperatives.

ABSENCE OF PRUDENT INTEREST RATE POLICIES

Cooperatives have access to huge deposits from public but there is no system to ensure prudent deployment of these funds. They have not taken advantage of the freedom given to them to decide on the interest rates on deposits and advances.

While deposits are accepted at unsustainably higher rates, loan pricing is done in imprudent manner without factoring in cost of funds. Many cooperatives are unable to generate enough revenue or surplus to sustain their operations and boost capital formation on account of very thin or negative margins. As a result, there are frequent demands for recapitalisation which causes pressure on the limited resources of governments which could have been used for productive investments benefitting the society in large.

LACK OF HR POLICIES AND PROFESSIONALISM

The quality of human resources is an important determinant of the success of any organization. This aspect, however, has not received due importance in the cooperative institutions. The cooperative banks are headed by a committee of elected members who are not professionals and do not possess sound knowledge in banking functions. The Committee takes crucial business decisions involving sanction of loans, investments, interest rates on deposits and loans, *etc.* which require a minimum degree of skill and expertise. Often, the role of Chief Executive Officers in these areas is minimal.

Cooperatives do not have well defined capacity building and HR policies in crucial areas like recruitment, placement, training, career progression, succession planning, *etc.* Recruitments are done without any objective and systematic manpower assessment. All these have led to inefficiency and lower productivity. Though there is a system of training in place in many cooperative banks, there is no need-assessment to align training with the current and future staff requirements. Training programmes should be designed to achieve skill upgradation in areas related to audit and expenditure, management and aptitude development. It is also necessary to keep the staff sufficiently motivated through periodic job rotation, job enrichment and recognition of performance.

POOR RECOVERY PERFORMANCE

Frequent loan waiver announcements by governments aimed to garner electoral support have vitiated the credit discipline among the borrowers and affected the recovery atmosphere. The resultant weak finances, growing NPAs and poor resource base have contributed to the declining performance of the cooperatives, particularly at the grass root level. Since, these grass root institutions depend on liquidity support from higher financing agencies like State and Central Cooperative Banks, non-payment of dues impairs the financial health of the entire chain.

INADEQUATE INTERNAL CONTROL AND AUDIT

Cooperatives do not pay attention to implementation of adequate and foolproof internal checks and control. Balancing of books and reconciliation of entries are in arrears for unduly long periods. As a result, there is frequent incidence of frauds involving the staff who game the system. Very often, these cases are held up due to protracted litigation in Courts of law and the guilty are not brought to book .There is absence of effective audit mechanism. There are delays in the conduct of audits and submission of reports. Quality of audit leaves much to be desired in as much as it fails to address the gaps in the systems and processes and improve efficiency of operations of the cooperatives.

LACK OF MEMBER PARTICIPATION

The cooperative structure should be member-driven. However, members having a voting right do not take active part or show interest in the affairs of the cooperatives since, the control and management is vested in a few members .Besides, depositors, whose money is intermediated by the cooperatives, have no voting right or any say in the management.

NOT KEEPING PACE WITH CHANGES

Cooperatives have been unable to adapt themselves to the rapid pace of changes in the financial sector. Cooperatives have lagged behind in designing of new products and services and in adoption of technology and advanced management practices that have changed the face of financial sector in the post economic-liberalization era. As a result, they are unable to cope with the stiff competition posed by other banks in rural finance and retain their market share.

DUALITY OF CONTROL

Under the Constitution, 'Cooperation' is a state subject governed by the respective State Cooperative Societies Acts. Registration, incorporation, management, election, and audit are governed by the State Acts. Some aspects relating to banking activities are regulated and supervised by the Reserve Bank of India / NABARD under the Banking Regulation Act, 1949 (As Applicable to Cooperative Societies). There is an urgent need to remove the overlapping controls and endowing functional autonomy and operational freedom to cooperatives. Banking functions should be brought completely under the Banking Regulation Act. The provisions of the Banking Regulation Act should override the provisions of the State Acts/bye-laws/rules which run counter to it. This will lead to clear demarcation of the areas of activities of cooperative banks.

NEED FOR REVIVAL OF THE COOPERATIVE CREDIT INSTITUTIONS

It is pertinent to mention that in the first place, India is a country with a population of more than 110 crore, of which around 70 crore reside in a little over 6 lakh villages. As there are a little over ninety two thousand PACS in the country, every 8th village, on an average, has an existing cooperative credit outlet. It is in this context that the role of cooperatives assumes importance. A number of Committees were set up to suggest reforms in the sector. Based on the recommendations of the Vaidyanathan Committee, the Government of India rolled out in January 2006, a package for revival of the Short-Term Rural Co-operative Credit Structure.

The Task Force also suggested wide-ranging reforms in the governance and management of STCCS including crucial amendments to the respective State Cooperative Societies Acts that were to precede the recommended one-time capitalization jointly by the Central government, the state governments and the STCCS of the state itself. Twenty-five state governments signed the MoU with Government of India and NABARD to participate in and implement the package. As on December 2012, 21 states had amended the respective State Cooperative Societies Act.

An amount of ₹ 9,002 crore was released by NABARD as the Government of India's share, while the state government released ₹ 856 crore as their share for recapitalisation of 53,202 eligible PACS in 17 States. Recapitalisation assistance could not be released in many cases as the states did not complete all the necessary benchmark activities within the stipulated period.

LICENSING OF RURAL COOPERATIVE BANKS

The Committee on Financial Sector Assessment set up by the Government of India in September 2006, under the Chairmanship of Dr. Rakesh Mohan, looked into the financial health of all the banks including the cooperative banks and made recommendations for improvement of financial health and systems for attaining and maintaining financial stability.

Table. Progress in Licensing of State and Central Cooperative Banks.

Position as on	StCBs			DCCBs		
	Licensed	Unlicensed	Total	Licensed	Unlicensed	Total
31 March 2006	13	18	31	73	298	371
31 March 2009	14	17	31	75	296	371
31 March 2012	30	01	31	329	42	371
30 September 2012	31	-	31	330	41	371
30 November 2012	31	-	31	345	26	371
31 March 2013	31	-	31	348	23	371
31 December 2013	32*	-	32	348	23	371

Note:

* The Jharkhand State Coop. Bank was issued license on 26 August 2013

A major recommendation of the Committee was to prohibit unlicensed banks from functioning beyond March 2012.

As there were a large number of cooperative banks (17 out of 31 State Cooperative Banks and 296 out of 371 District Central Cooperative banks) functioning without license, the Reserve Bank of India relaxed the licensing norms to grant license to these banks. The relaxed norms were

- Capital to Risk-weighted Assets Ratio (CRAR) – Minimum 4 per cent, and
- Compliance with CRR and SLR for the last one year (default on one or two occasions were permitted).

Consequent upon relaxation of licensing norms, 50 banks qualified for issue of license reducing the number of unlicensed banks from 73 as on February 29,

2012 to 43 as on March 31, 2012 and further to 23 as on June 30, 2013.Recent Policy Initiatives

IMPLEMENTATION OF CORE BANKING SOLUTION

The Reserve Bank of India has granted permission to StCBs and CCBs to participate in the payment system and offer RTGS/ECS/NEFT facilities to their customers/members. Since, IT infrastructure is the prerequisite for offering these services, Core Banking Solution (CBS) is being implemented in StCBs and CCBs. Several banks have joined the NABARD assisted project and made significant progress in implementation of CBS. Some banks have undertaken implementation of CBS on their own. Full implementation is expected to be achieved by March 31, 2014.

Table. Progress in Implementation of CBS as at January 31, 2014.

No. of banks		CBS completed		CBS in in progress		Yet to commence	
StCBs	DCCBs	StCBs	DCCBs	StCBs	DCCBs	StCBs	DCCBs
32	371	28	294	4	55	-	22

PRUDENTIAL NORMS ON CAPITAL ADEQUACY

Rural cooperative banks have been kept out of Capital to Risk-Weighted Assets Ratio (CRAR) framework for a long time on the grounds that there is an in-built accretion to capital every time a loan is availed of by a member. In view of the high level of default characterizing the STCCS and the risk posed by higher concentration of loan portfolios of these banks in a single sector with defined areas of operation, the Vaidyanathan Task Force recommended for bringing all tiers of the STCCS under the CRAR framework. The Reserve Bank of India recently reviewed the position of CRAR in StCBs and CCBs and has introduced the minimum CRAR requirement for these banks. A roadmap has been laid down for achieving CRAR of 7per cent by March 31, 2015 and 9per cent by March 31, 2017.

Since, these banks have limited avenues of mobilizing additional capital resources, permission has been granted for issue of Long Term (Subordinated) Deposits and Innovative Perpetual Debt Instruments. State / Central Co-operative Banks (StCBs / CCBs) are now allowed to issue Long Term (Subordinated) Deposits (LTD) with the prior permission of the respective Registrar (RCS) granted in consultation with the Reserve Bank. LTD can be issued to members and non-members, including those outside the area of operations of the StCBs / CCBs concerned. There is no prohibition on existing shareholders subscribing to LTD. The amounts raised through LTD will be eligible to be treated as lower Tier II capital, subject to certain conditions. State / Central Co-operative Banks are also allowed to issue Innovative Perpetual

Debt Instruments (Innovative Instruments). The instruments may be issued as bonds or debentures to qualify for inclusion as Tier I Capital for capital adequacy purposes. Banks are required to obtain prior approval of the Reserve Bank of India, on a case-by-case basis, for issue of innovative instruments.

Innovative instruments shall not exceed 15 per cent of total Tier I capital. Innovative instruments in excess of the above limits shall be eligible for inclusion under Tier II, subject to limits prescribed for Tier II capital. However, investors' rights and obligations would remain unchanged.

MEMBERSHIP OF CREDIT INFORMATION COMPANIES

Under the Credit Information Companies (Regulation) Act, 2005, all co-operative banks are credit institutions and every credit institution has to be a member of at least one Credit Information Company (CIC). State and Central cooperative banks are required to take membership of at least one CIC.

In developing economies, lending, especially to retail, has been made possible by credit information companies, which perform the vital task of collating and distributing reliable credit information to the specified users. Reports from CICs contain information about the payment behaviour of consumers and commercial entities, including data on timely fulfillment of or delinquency in meeting the financial obligations. It helps credit institutions in making an informed decision about a credit applicant.

STATE COOPERATIVE BANKS – GUIDING ROLE

State Cooperative Banks are the apex level institutions in the federal structure of STCCS. They accept huge deposits from the affiliated / member societies and also from non-members. They provide liquidity support and financial accommodation to the member societies down the chain. Large volume of funds in the form of governmental assistance under various welfare and subsidy schemes, grants and refinance from higher refinance agencies are channeled through state cooperative banks to the lower tier institutions. Hence, it is the responsibility of the state cooperative banks to provide guidance and direction to the member societies and ensure that the valuable resources are not frittered away and diverted to investments in risky assets. Weaknesses in lower tier institutions would eventually move up the chain and bring down the entire edifice of the federal cooperative credit structure. Even strong state cooperative banks would face threat to their survival if the structural weaknesses are not addressed in a timely manner.

State cooperative banks have a larger role in the development of the state economy. They should rise above the self-interest and work for serving the economic interests of the society at large. It need not be emphasized that these banks should keep their own house in order and conduct themselves with unquestionable integrity and social responsibility.

ROLE OF COOPERATIVES IN PROVIDING AGRICULTURAL CREDIT

The main players in the field of agricultural credit in the formal sector include the commercial banks, the regional rural banks (RRBs), and the rural cooperatives. The rural credit cooperatives in the country are in an impaired state. Several factors have led to the impairment of the Cooperative Credit Structure, but it would be advisable to understand the magnitude of the problem first. The cooperatives once dominated the rural credit market in the institutional segment (with a share of around 65 per cent, going by the All India Debt and Investment Survey 1991), but now have a significantly smaller role.

Data for the past decade indicates a fall in the share of cooperatives in the rural credit market, from around 62 per cent in 1992-93 to about 34 per cent in 2002-2003 inspite of an increase of just under 10per cent per annum in the absolute disbursement on a compounded annual basis.

Two trends emerge from the overall flow of credit to agriculture from the commercial banking sector. The number of rural branches of commercial banks has gone down marginally as part of the branch rationalisation programme. The second trend is that even though the commercial banks almost meet their targets for lending to the priority sector, they have moved more towards larger customers. The average size of direct loans to agriculture in the portfolio of the commercial banks was ₹ 13,500 in 1997, and is ₹ 31,585 now. The average size of loans of the PACS, in comparison, is currently only ₹ 6,640 per borrower, according to the data tabulated overleaf.

Thus, in a country predominated by small or marginal land holdings, the reach of the cooperative system is much deeper than the other institutional arrangements in the rural areas.

Notwithstanding the falling share of cooperatives in the overall share of institutional credit practically in all States, it was found that in States like Gujarat, Maharashtra, Haryana, Madhya Pradesh, Chhattisgarh, Orissa and Rajasthan, the share of cooperatives in institutional credit is currently 50 per cent or more. In States like Bihar, Jharkhand, Himachal Pradesh and Assam, their share is negligible.

The traditional banking system, the systems and procedures of which are actually designed for the urban industrial and business financing, has limitations in reaching out to the last mile. The exposures of the banks for this segment have risen, but, the rates for defaults in repayment have also gone up. Most often, this happened because banks have not applied appropriate methods for banking with the poor, by keeping in touch with the customers and applying social collaterals. Banks have traditionally worked on documentation related appraisals, rather than on trust and production related appraisals. The client group, however, needs much more support than what the banks currently provide. By implication, we need to necessarily look to the cooperative sector for delivering credit to small and marginal farmers, and those who have little or no productive assets. It is, therefore, imperative that the cooperative sector, particularly at the primary level, be revived on a priority basis.

IMPAIRMENT OF GOVERNANCE

World over, cooperative credit structures have been based on the concept of mutuality, with thrift and credit functions going hand in hand. But, in India, the structure has largely been focused on credit. The primary level cooperatives, therefore, have traditionally been agencies for credit dispensation. Because of this characteristic at the base level, the upper tiers were created to ensure that the lower tiers get refinance. The structure is, therefore, driven by borrowers at all levels, which creates a serious conflict of interest. A solution is to aggressively advocate conversion of pure credit to thrift cum credit cooperatives. Such societies would not only increase the financial stakes of the members in the system, but also factor in natural incentives for better governance.

The impairment in governance is deep and is represented by the composition of the boards of directors of the cooperatives and the reporting systems. Because of the structural ordering, the lower tiers are managed by the higher tiers in varying degrees of detail in different States. In almost all States, the function of conducting elections for the cooperative structure is vested with the State Government. Similarly, the function of auditing is also vested with a State-run audit system. By implication, the cooperatives lose their right to self-governance and have to look up to the State constantly for several of the functions that naturally fall in the domain of the general body and the Board of Directors. Some pointers on the governance systems are highlighted below:

- No elections have been held in the CCS units across all tiers for long (10 years or more) in three States
- Boards of nine out of 30 SCBs and 134 out of 368 DCCBs have been superseded
- Most State Governments combine the roles of Dominant Shareholder, Manager, Regulator and concurrent Supervisor and Auditor
- The Department headed by the Registrar of Societies (RCS) can and does, influence administrative matters. The interference is in the form of supersession of Boards, appointment of administrators and assuming powers to approve staffing patterns, recruitment, emoluments, asset purchase pattern *etc.*
- The Department also interferes in financial matters in various forms, like direction on interest rates, interference in loan decisions, announcement of waivers, and direct or indirect pressure on non-recovery of loans

- The impairment of the governance structure is also because of politicisation of these institutions, reflected in the fact that directors on Boards of Cooperative Banks are involved in active politics either at the State, District, and Taluka level. Data on political background of Directors on Boards of SCBs.
- Audit is pending in at least 15 per cent of the PACS for more than a year. This is a optimistic estimate. Audits are more regular in the upper tiers. Apart from delays, the quality of audit needs to be examined carefully. As the State machinery is involved in conducting audits, those actually conducting audits may not be professionally trained to audit financial cooperatives. It is therefore, doubtful that they are able to understand and comment on the reporting of the actual financial position of PACS.
- Audit at the higher tiers are done in a relatively efficient manner, the income recognition and provisioning norms are more standardized and therefore, the accounting data from the higher tiers could be assumed to be relatively more reliable.

Nevertheless, the audit classification of some banks in some

States seems to suggest that the audited results do not depict a true and fair position of the banks concerned. While there are issues of internal governance that are a cause of concern, we also have to remember that even the external regulation and supervision for the structure are not as stringent as it is for the commercial banking structure. In particular the following aspects are to be considered:

- Primary agricultural societies (PACS) are excluded from the scope of the BR Act,1949
- The minimum capital requirement is only ₹ 1 lakh for banks
- The cash reserve ratio (CRR) requirements are lower than that for commercial banks
- The Capital at Risk Weighted Asset Ratio (CRAR) norms have not been prescribed even for SCBs and DCCBs
- All CCS units are, however, subject to submission of regular returns on their financial status and operations, the compliance of which is weak
- The cooperative banks are open to periodic inspection by NABARD. The compliance with the supervision findings and regulations is, however, weak.

The central regulatory authority (the RBI) is naturally concerned at its inability to ensure that financial institutions comply with even the relatively diluted prudential norms applicable to them and to enforce punitive measures against banks that are in poor and deteriorating financial health. The RBI's plight may be attributed to three primary reasons, of which dual control of cooperative banks by the RBI and the State Governments, is one. The ambiguities on the precise jurisdiction of powers between the two, and the reluctance of the State Governments to enforce disciplinary sanctions by the RBI, are others. Attempts to change the law (through the Banking Regulation Amendment bill) have failed.

The States (and in some cases the Union Government) have not helped the regulatory authority. On the contrary, their actions (*e.g.*, waiver of loans in 1989 by the Union Government, periodic waivers of interest and principal by the State Governments, delay in payments by the State Governments on promises made, their formal or informal instructions to delay or dilute loan recovery, and their unwillingness to facilitate recoveries under the Revenue Recovery Act) have contributed to an atmosphere, that encourages defaults in payment and worse.

IMPAIRMENT IN MANAGEMENT

The impairment in the management of the rural cooperatives is a direct result of the impairment in governance. The various forms of interference of State Governments include deputation of officials to top positions in many banks, setting up common cadres for senior positions in cooperatives across tiers, determination of staffing pattern, and interference in the operational decisions of the cooperatives. The Task Force has sought to collect information on areas in which the state governments are involved in the operational aspects of cooperative banks. The details that the Task Force has been able to collect from the Regional Offices of NABARD.

The impairment in management is also owing to the following additional factors:

- Managers of PACS in several States are drawn from a common staff pool who do not feel accountable to the PACS. Remuneration often is without reference to business level or results.
- A generally ageing staff profile characterised by inadequate professional qualifications and low levels of training.
- Delineation of Governance and management functions are unclear and the boards take up issues at operational level, thereby losing sight of the long term strategic issues
- Poor housekeeping, weak internal controls and systems.

The cumulative result is that members, who are mostly borrowers, have little or no sense of stake in the cooperatives, or any accountability in ensuring prudent management of funds. On the contrary, government policies (loan and interest waivers, delaying recoveries, the fact that loans carry State guarantees) encourage them to presume that they can with impunity, delay or even fail to meet their repayment obligations. Boards of management and their functionaries are not held accountable for laxity in granting and monitoring loans, poor quality of loan portfolios, high default rates and non-performing assets (NPAs) and their adverse effects on the financial health and viability of the societies.

AGRICULTURE CREDIT IN INDIA

Agriculture credit is an important prerequisite for agricultural growth. Agricultural policies have been reviewed from time to time to provide adequate and timely availability of finance to this sector. Rural credit system assumes importance because for most of the Indian rural families, savings are inadequate to finance farmingand other economic activities.

This coupled with the lack of simultaneity between income realization and expenditure and lumpiness of agricultural capital investments.

The institutional credit system is critical for agricultural development and its role has further increased in the liberalized economic environment. In India a multi-agency approach comprising co-operative banks, scheduled commercial banks and regional rural banks (RRBs) has been followed to allow credit to agricultural sector.

TYPES OF AGRICULTURE CREDIT

The agriculture credit can be classified on the basis of:

- According to Tenure of Agricultural Credit *i.e.* the credit requirement based on the time-period of loans.It can of three types:
  - *Short-Term:* It refers to the loans required for meeting the short-term requirements of the cultivators.These loans are generally for a period not exceeding and repaid after the harvest. For example loans required for the purchase of fertilizers, HYV seed, for meeting expense on religious or social ceremonies *etc.*
  - *Medium-Term:* These loans are for a period up to 5 years. These are the financial requirements to make improvements on land, buying cattle or agricultural equipments, digging up of canals *etc.*
  - *Long-Term:* These loans are for a period of more than 5 years and are generally required to buy additional land or tractor or making permanent improvements on land.
- According to Purpose of Agriculture Credit: The agriculture credit on the basis of purpose for which the credit is used can be of two types:
  - *Productive:* Productive loans are the loans that are related to agricultural production and economically justified. For example purchase of tractor, land, seeds *etc.*
  - *Unproductive:* Unproductive credit are used for personal consumption and unrelated to productive activity for example loans for expenditure on marriages, religious ceremonies *etc.*

SOURCE OF AGRICULTURAL CREDIT IN INDIA

There are two broad sources of agricultural credit in India:

- Non-Institutional Sources
- Institutional Sources

Non-Institutional Sources

The non-institutional finance forms an important source of rural credit in India,constituting around 40 percent of total credit in India. The interest charged by the non-institutional lenders is usually very high. The land or other assets are kept as collateral. The important sources of non-institutional credit are as follows:

- *Money-Lenders:* Money-lending has been the widely prevalent profession in the rural areas. The money-lenders charge huge rate of interest and mortgage the property of the cultivators and in some cases even the peasants and members of his family are kept as collateral.
- Other Private Sources:
  - *Traders, landlords and commission agents:* The agents give credit on the hypothecation of crops which when harvested is used to repay loans.
  - *Credit from relatives:* These credits are generally used for meeting personal expenditure.

Institutional Sources

The general policy on agricultural credit has been one of progressiveinstitutionalization aimed at providing timely and adequate credit to farmers for increasing agricultural production and productivity. Providing better access to institutional credit for the small and marginal farmersand other weaker sections to enable them to adopt modern technology and improved agricultural practices has been a major thrust of the policy. National Bank for Agriculture and Rural Development (NABARD) is an apex institution established in 1982 for rural credit in India. It doesn't directly finance farmers and other rural people. It grants assistance to them through the institutions described as follows:

RURAL CO-OPERATIVE CREDIT INSTITUTIONS

Rural Credit cooperatives are the oldest and most extensive form of rural institutional financing in India. Themajor thrust of these cooperatives in the area of agricultural credit is the prevention of exploitation of the peasants by moneylenders. The rural credit cooperatives may be further divided into short-term credit cooperatives and long-term credit cooperatives.

Short-term Credit Cooperatives

The short-term credit cooperatives provide short-term rural credit and are based on a three-tier structure2 as follows:

Primary Agricultural Credit Societies (PACs)

These are organised at the village level. These societies generally advance loans only for productive purposes. The main objective of a PACS is to raise capital for the purpose of giving loans and supporting the essential activities of the members such as supply of agricultural inputs at cheap price, improving irrigation on land owned by members, encourage various income-augmenting activities such as horticulture, animal husbandry, poultry *etc.* In India, around 99.5 percent of villages are covered by PACs.

District Central Cooperative Banks

These cooperatives are organised at the district level. The PACS are affiliated to the District Central Co-operative Banks (DCCBs). DCCBs coordinate the activities of district central financing agencies, organise credit for PACs and carry out banking business.

State Co-Operative Banks

The DCCBs are affiliated to State Co-operative Banks (SCBs), which coordinate the activities of DCCBs, organise provision of finance for credit worthy farmers, carry out banking business and act as leader of the Co-operatives in the States.

Long-term Credit Cooperatives

Long-term credit Cooperatives: These cooperatives meet long-term credit of the farmers and are organised at two levels:

Primary Co-Operative Agriculture and Rural Development Banks

These banks operate at the village level as an independent unit.

State Co-Operative Agriculture and Rural Development Banks:

These banks operate at state level through their branches in different villages.

COMMERCIAL BANKS

Commercial Banks(CBs) provide rural credit by establishing their branches in the rural areas. The share of commercial banks in rural credit was very meagre till 1969. The All India Rural Credit Review Committee (1969) recommended multi agency approach to the rural and especially agricultural credit. It suggested the increasing role of the CBs in providing agricultural credit. Further, under the Social Control Policy introduced in 1967 andsubsequently the nationalization of 14 major CBs in 1969 (followed by another six banks in 1980), CBs have been given a special responsibility to set up their advances for agricultural and allied activities in the country.The major expansion of rural branches took place and CBs introduced Lead Bank scheme and district credit plans for rural areas. Banks were asked to lend 18 percent of their total advances to agriculture within thequota of 40 percent of priority sector lending. This expansion of rural credit remained till the late 1980s.However, during late 80's, CBs suffered huge losses due to waiving of agricultural loans by the government.The financial liberalization process with the adoption of Narasimham Committee report in 1993 hasnecessitated the banks to focus on profitability and adopt prudential norms. The proportion of bank credit to rural areas especially small borrowers has come down steadily.

REGIONAL RURAL BANKS (RRBS)

RRBs are the specialised banks established under RRB Act, 1976 to cater to the needs of the rural poor. RRBs are set-up as rural-oriented commercial banks with the low cost profile of cooperatives but with the professional discipline and modern outlook of commercial banks. Between 1975 and 1987, 196 RRBs were established with over 14,000 branches. As a result of the amalgamation, the number of RRBs was reduced from 196 to 133 as on 31 March, 2006 and to 96 as on 30 April 2007. RRBs covcred 525 out of 605 districts as on 31 March 2006. After amalgamation, RRBs have become quite large covering most parts of the State.

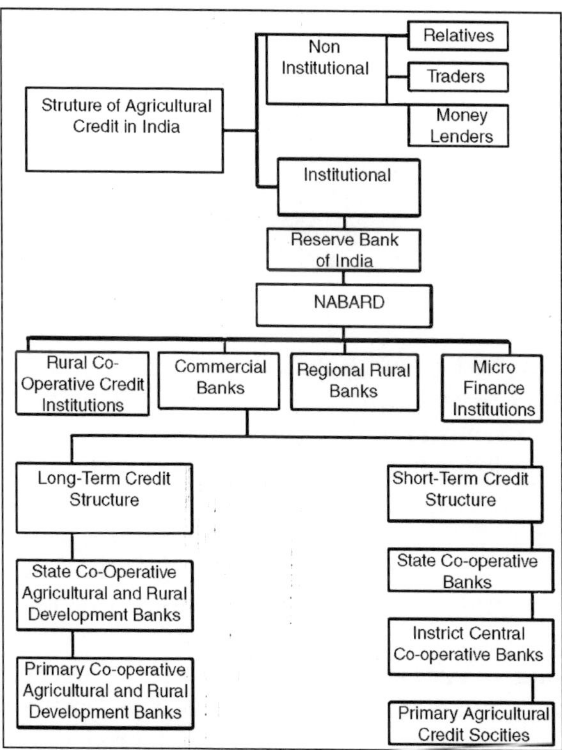

Fig. Structure of Agricultural Credit System in India

Increasedcoverage of districts by RRBs makes them an important segment of the Rural Financial Institutions (RFI). The branch network of RRBs in the rural area form around 43 per cent of the total rural branches of commercial banks. A large number of branches of RRBs were opened in the un-banked or under-banked areas providing services to the interior and far-flung areas of the country. RRBs primarily cover small and marginal farmers,landless laborers, rural artisans, small traders and other weaker sections of the rural community. However,even after so many years, the market share of RRBs in rural credit remained low and have suffered huge losses.In recent years Government has initiated reform process to improve the functioning of RRBs.

Efforts are made to increase the capital base and investible fund of these banks. The financial support is provided to improve training, technology development including computerization in these banks. The structural consolidation process has been initiated by amalgamating these banks.

This has resulted into pooling of resources including experienced workforce, common marketing efforts and thus better customer services.Further, RRBs have been able to derive the benefits of increased area of operations and enhanced credit exposure. These measures have provided remarkable improvements in the financial performance of RRBs. The number of RRBs incurring losses and the levels of non-performing assets has reduced dramatically.

MICRO FINANCE INSTITUTIONS (MFIS)

Banks offer concessional interest rates for the rural credit. However; small farmers are unable to access thembecause of borrower-unfriendly products and procedures, inflexibility and delay, and high transaction costs,both legitimate and illegal.

Thus, Non-Government Organisations (NGOs) are providing alternative means to enhance access to credit by the poor since, mid-70's. After pioneering efforts by organisations like SEWA,MYRADA, PRADAN and CDF, in 1992 the RBI and NABARD encouraged commercial banks to link up with NGOs to establish and finance self-help groups (SHGs) of the poor.

The RBI has included financing of SHGs under priority sector lending. At present, there are three groups of SHGs *viz.* SHGs formed and financed by the banks (20 percent); SHGs formed by other formal agencies but financed by banks; SHGs financed by banks using NGOs and other agencies (8 percent).These institutions provide small loans to the poor at low interest rates without collateral.

The experience of micro-finance scheme in India suggests that i) It is the cost effective way of financing the rural poor; ii) The repayment rate of SHGs is more than 95 percent due to peer pressure; iii) It reduces transaction costs of borrowers as well as lenders; iv) It inculcates the habit of thrift among members and provide timely credit.

TRENDS IN AGRICULTURAL CREDIT

Over time, spectacular progress has been achieved in terms of the scale and outreach of institutional framework for agricultural credit. Some of the major discernible trends are as follows:

INCREASING DEPENDENCE ON INSTITUTIONAL CREDIT

One of the major achievements in the post-independent India has been the widening of the spread of institutional machinery for credit and decline in the role of non-institutional sources.

The share of institutional credit, which was little over 7 per cent in 1951, increased manifold to over 66 per cent in 1991, reflecting a remarkable decline in the share of non-institutional credit from around 93 per cent to about 31 per cent during the same period. However, the latest NSSO Survey reveals that the share of non-institutional credit has taken a reverse swing which is a cause of concern .

Table. Relative Share of Borrowing of Cultivator Households from Different Sources.

(Figures are in %)

Sources Credit	1951	1961	1971	1981	1991	2002
Non-Institutional	92.7	81.3	68.3	36.8	30.6	38.9
Money Lenders	69.7	49.2	36.1	16.1	17.5	26.8
Institutional	7.3	18.7	31.7	63.2	66.3	61.1
Cooperatives	3.3	2.6	22.0	29.8	23.6	30.2
Societies / Banks						
Commercial Banks	0.9	0.6	2.4	28.8	35.2	26.3
Unspecified	–	–	–	–	3.1	–
Total	**100.0**	**100.0**	**100.0**	**100.0**	**100.0**	**100.0**

TRENDS OF OVERALL INSTITUTIONAL CREDIT FLOW

Over time, the flow of credit to agriculture and rural sector has expanded impressively . The ground level credit flow had registered an increase from ₹ 1675 crore in 1975-96 to ₹86891 crore in 2003-04 and further to ₹203297 crore in 2005-06. This rate of growth was even higher than the growth rate of GrossDomestic Product (GDP) originating in agriculture. Despite this growth, the credit needs of agriculture have not been met fully and overwhelming numbers of farm households have not been able to borrow from institutional sources. While short-term credit has remained the dominant component of total credit, its relative importance declined from 70.3 per cent in 1975-76 to 58.1 per cent in 2006-07.

Table. Institutional Credit Flow to Agriculture Sector (₹ Crore)

Agency	1975-76	1983-84	1993-94	2001-02	2002-03
Cooperatives	1186	2938	10117	23524	23636
Regional Rural Banks	2	263	977	4854	6070
Scheduled Commercial Banks	405	1885	5400	33587	39774
Other Agencies	82	185	0	80	80
Total Credit	1675	5244	16494	62045	69560
Agency	**2003-04**	**2004-05**	**2005-06**	**2006-07**	**2007-08**
Cooperatives	26,875	31231	39404	42480	33070
Regional Rural Banks	7581	12404	15223	20435	15925
Scheduled Commercial Banks	52441	81481	125477	140382	88765
Other Agencies	84	193	382	NA	NA
Total Credit	86981	125309	180486	203297	137760

Note * Up to November 2007. Source: All India Debt and Investment Survey and NSSO.

AGENCY-WISE CREDIT FLOW

The analysis of agency wise credit flow indicates that the cooperative bank were the major source of agriculture credit in 1975-76 constituting around 71 percent of the total ground level credit flow followed by commercial banks at 24.2 percent and regional rural banks at 4.9 percent . Though cooperative banks had dominated agriculture credit supply till the early reform period, commercial banks and RRBs recorded impressive growth rates.

As a result, in 2006-07, the share of cooperative banks in the total institutional credit flow receded to 20.1 percent and that of commercial banks advanced to 69.1 percent. Although the quantum of disbursement from cooperative banks increased, it could not keep pace with commercial banks in enhancingcredit flow due to several reasons including its poor financial health, dual control, lack of internal controls and corporate governance norms and excessive dependence on other financial institutions.

Although the share of cooperative credit is now much lower than that of commercial banks, the reach of cooperative credit societies is much wider. Cooperative credit societies have more than twice the number of rural outlets and four times more accounts than those of scheduled commercial banks and RRBs put together

Low recovery rates and mounting overdues have clogged the process of recycling of credit by cooperatives,impaired their ability to avail of refinance facilities from (NABARD), increased transaction costs and more importantly, have deprived potential borrowers of the opportunity to avail of credit facilities from the cooperatives. As a result, cooperatives have been losing their capacity to meet the growing credit needs of agriculture.

SIZE-WISE CREDIT FLOW

Despite impressive growth in direct credit to farmers from the scheduled commercial banks between 1991-92and 2003-04, contrary to expectation, credit disbursement to small and marginal farmers has not been encouraging. Though the number of accounts increased for small farmers yet the credit flow favoured the richer farmers.

REGION-WISE CREDIT FLOW

While analyzing the pattern of credit flow, it is observed that the proportions of bank deposits and credit shares have moved in favour of the South, West and North regions. While the share of loans in the total disbursement of credit for agriculture and allied activities were the maximum for the South region, it was the minimum for North-east region.

RECENT INITIATIVES

Government of India and Reserve Bank of India: In order to increase credit flow to the agriculture sector, the policy of doubling of agricultural credit in three years was introduced in 2004-05. In order to expand the outreach of the banking services,banks made available basic banking 'no-frills' account with low or nil minimum balances as well as low or no charges in 2005. The regional rural banks were also specifically advisedto allow limited overdraft facilities in 'no-frills' accounts without any collateral or linkage to any purpose.

National Agricultural Insurance Scheme (NAIS)

NAIS is implemented since, Rabi 1999-2000 season with the objectives to provide insurance coverage and financial support to the farmers in the event of failure of any of the notified crops as a result of natural calamities, pests and diseases and to encourage the farmers to adopt progressive farming practices, high value in-puts and higher technology in agriculture and to help stabilize farm incomes particularly in disaster years.

Government of India & NABARD

Rural Infrastructure Development Fund

RIDF was established in 1995-96 with a corpus of ₹ 2,000crores with the major objective of providing funds to state governments and state owned corporations to develop rural infrastructure such as rural roads, rural bridges, irrigation works, soil conservation, flood protection,drinking water, infrastructure for rural education *etc.* The total corpus of RIDF till 2007-08 (RIDF-I to RIDF-XIII) amounted to ₹72, 000 crores with 2008-09 budget further adding the amount of RIDF XIV of ₹ 14,000 crores to this corpus. The total sanctions and disbursements as on 30 June 2007 aggregated ₹ 61312.27 crores and ₹38581.82 crores respectively.

Micro Finance Innovations

The credit accessibility for the poor from conventional banking is limited due to lack of collaterals and information. Micro finance has emerged as an alternative financial vehicle that provides micro credit or small loans granted to the poor without any collateral. These loans are provided through micro finance institutions (MFIs). NABARD plays a key role in developing the MFIs by providing them refinance facilityat low interest rates.

Kisan Credit Card Schemes

The kisan credit cards (KCC) scheme was introduced in 1998-99 to facilitate short-term credit to farmers. Each farmer is given with a kisan credit card and a pass book for providing revolving cash credit facilities. NABARD provide refinance facility to commercial banks and cooperatives to provide credit under this scheme.

Refinance under Swarnajayanti Gram Swarozgar Yojna (SGSY)

NABARD provides refinance facility to institutions that support SGSY.

Co-operative Development Fund

NABARD has set up the cooperative development fund in 1993 with objective of strengthening the co-operative credit institutions in the areas of resource mobilization, recovery position *etc.* the assistance is provided to cooperatives by way of soft loans or grants.

WEAKNESSES IN RURAL CREDIT STRUCTURE

Overemphasis of Monetary Credit

The rural credit institutions have given overemphasis on the financial assistance to the cultivators. While the finance is very important factor but it should be complemented with the extension of services in form of guidance, expertise and counseling on agricultural issues.

Multiplicity of Institutions

The rural credit structure is based on multi agency credit system whereby there exist numerous organisations providing similar kind of financial services.

There is a lack of coordination in the system and the commercial viability is adversely affected in this scenario.

Lack of Motivation

In order to fill the gap that occurred due to the failure of rural cooperative societies Government gave increasing role to the commercial banks. However, commercial banks lack the desired skillsand expertise in the agro-credit. The banks have enough financial resources but the service consultancy is not available.

Thus, there is a failure to provide complete package of assistance to the farmers. Further, financial sector reforms have put pressure on banks to improve their financial position and so these banks are now concentrating on selected clientele of large borrowers.

Financial Exclusion

Despite of a large network of the institutional credit system, it has not been able to adequately penetrate the informal rural financial markets and the non-institutional sources continue to play adominant role in purveying the credit needs of the people residing in rural areas.

The results of the All-India Debt and Investment Survey (AIDIS, 2002) also indicate that the share of the non-institutional sources, in the total credit of the cultivator households, had increased from 30.6 percent in 1991 to 38.9 percent in 2002.

High Interest Rates

The rate of interest charged by rural financial institutions (RFIs) from farmers continues to be considerably higher than those charged by financial institutions from urban consumers. The owners of small or marginal farms, which are non-viable or viable at the margin, and self-employed in the informal sector, cannot afford to bear the level of interest charged by RFIs.

Procedural Delays

There is a problem of considerable delays in processing of loan applications and collaterals. Thus farmers shy away from institutional financing and increase their dependency upon non-institutional sources.

Poor Recoveries

Banks are shying away from rural financing mainly because of poor recoveries which is inflicting the system. It is ironical that the recoveries position is adverse amongst rich farmers than amongst the small farmers. The political decisions of waiving off loans are further putting pressures on the financial system.

SUGGESTIONS FOR IMPROVING INSTITUTIONAL RURAL CREDIT SYSTEM

Financial Discipline to Improve Recovery

A national consensus among political parties should be evolvedfor not politicizing the RFIs and resist from announcement of loan or interest waiver schemes and giving calls for not repaying the institutional loans. However, given the risk involved in the agriculture credit the recovery system should be flexible and humane.

Revamping the Cooperative Credit Structure

The Cooperative Credit Structure should be strengthened to make use of its wider reach. These have to be recapitalised so as to provide funds for improving their financial positions. There is a need of capacity building, human resource development, institutional restructuring to ensure democratic functioning, and improving the regulatory regime to empower the Reserve Bank of India (RBI) to enforce prudent financial management.

Better Physical, Social and Economic Infrastructure

The long term policy framework needs to be designedto improve infrastructure facilities so as to boost rural economic growth. This requires increased public expenditure on social infrastructure (like education, availability of drinking water, health facilities), physical infrastructure (like roads, power) and economic infrastructure like (irrigation, modern agricultural techniques).These measures would help to improve the debt paying capacity of rural poor and provide greater opportunities to RFIs.

Financial cum Consultancy Approach

RFIs needs to provide extension services like consultancy about seeds, availability and use of modern inputs, marketing strategies etc to the cultivators so that a holistic package of assistance can be provided to them.

Group Approach to Lending

The lending to homogenous farmer's groups needs to be organized to improve credit delivery. This would help to improve recovery because of peer pressure. Further, group lending tends to be cost-effective. Involving NGOs or rural educated youths in organizing farmers or rural families in groups,scrutinizing applications, disbursement of loan and effecting recoveries would help RFIs in reducing lending costs.

Autonomy to RRBs

RRBs should be given more autonomy and flexibility in planning and lending policies, sothat their comparative advantage in rural lending is restored.

Greater Involvement of Micro Finance Organisations

The banks need to involve micro-finance agencies like SHGs, NGOs *etc.* and other grass root level financial intermediaries who have better understanding of the credit needs and recovery situations.

Technological Up Gradation

Technological improvements like computerization can be critical in building up a reliable credit information system and database on customers, reducing transaction costs and facilitating better pricing of risk, improving the efficiency of the financial system, and thereby increasing the access of un-banked rural people in an efficient manner.

Information Dissemination to Rural Poor

Credit counseling, awareness and financial education regarding the benefits of institutional financing are important for effective expansion of financial services in rural areas.To do this, banks may utilize the services of non-governmental organisations, village youth clubs, village panchayats, farmer clubs and self-help groups into confidence.

3

Agricultural Price Policy and Economic Development

INTRODUCTION

Agricultural Price Policy plays an important role in achieving growth and equity in the Indian economy in general, and the agriculture sector in particular. The major underlying objective of the Government's Price Policy is to protect both producers and consumers. Achieving food security at both the national and household levels is one of the major challenges in India today. Currently, the Food Security System and Price Policy basically consist of three instruments: Procurement Prices/Minimum Support Prices (MSPs), Buffer Stocks and the Public Distribution System (PDS). Agricultural Price Policy is one of the important instruments in achieving food security by improving production, employment and incomes of the farmers. There is a need to provide remunerative prices for farmers in order to maintain food security and increase the incomes of farmers. In India, the agriculture price policies and allied instruments were evolved in the pr-Independence era. The procurement and distribution of major food grains were started and statutory maximum prices were fixed, but were not strictly enforced. In the post-Independence era, the objective of achieving food security was linked with environment prices with a view to encourage higher investment and production. Though the Government decided to purchase food grains at fixed prices, if market prices fell precipitously, but till 1954 there was no sharp decline in food prices.

Importance of Price Policy

The prices of agricultural product fluctuate more quickly as compared to the industrial product. So these changes in prices affect the income, standard of living of the farmer and rural population. Even these also affect the trade of other goods.

Example:- In India and Pakistan if the prices of cotton or wheat falls any year, it also affects the trade and other business badly. The farmers aggregate demand falls, which affects the whole economy. The govt. in this situation interfere so that prices may not fluctuate beyond a particular limit. Because if the price of any particular product falls any year, then next year farmers never cultivate that product. It creates shortage next year in the market. So agricultural price policy has greater importance in the developing countries of the world.

Objectives of Price Policy

The objectives of agricultural price policy may differ from country to country. These depend upon the stages of development in a country. Anyhow, following are the important objects of agricultural price policy:

1. To meet the domestic consumption requirement govt. promotes balanced increase in production.
2. To provide price stability in the agricultural product.
3. To meet the national targets by the planners.
4. To provide the wheat to consumer to reasonable price.
5. To provide raw material to the industries at reasonable price.
6. To increase the production and exports of agricultural product.

IMPORTANCE OF PRICE POLICY

The price of agricultural product fluctuate more quickly as compared to the industrial product. So these changes in prices affect the income, standard of living of the farmer and rural population. Even these also affect the trade of other goods. *Example :* In India and Pakistan if the prices of cotton or wheat falls any year, it also affects the trade and other business badly. The farmers aggregate demand falls, which affects the whole economy.

The govt. in this situation interfere so that prices may not fluctuate beyond a particular limit. Because if the price of any particular product falls any year, then next year farmers never cultivate that product. It creates shortage next year in the market. So agricultural price policy has greater importance in the developing countries of the world.

Objectives of Price Policy

The objectives of agricultural price policy may differ from country to country. These depend upon the stages of development in a country. Anyhow, following are the important objects of agricultural price policy :

- To meet the domestic consumption requirement govt. promotes balanced increase in production.
- To provide price stability in the agricultural product.
- To meet the national targets by the planners.
- To provide the wheat to consumer to reasonable price.
- To provide raw material to the industries at reasonable price.
- To increase the production and exports of agricultural product.

METHODS OF ADOPTING AGRICULTURAL PRICE POLICY

To regulate the prices of product govt. taken the following measures :

Administrative Price

Govt. tries to maintain a favourable prices acceptable to both farmer and producer. But generally both the parties remain unhappy. If govt. increases the price of agricultural product, then consumer suffers. As govt. of increased the price of wheat to encourage the farmer, public criticized the govt. badly. On other side if goods are prices low then farmers suffers and production is affected adversely. Govt. tries to protect the interest of the both parties.

The administrative prices consists upon the following the prices :

- *Support Prices :-* Every year govt. fixes the support prices of important crops. The prices are announced before the sowing time, to encourage the farmers.
- *Issue Prices :-* To protect the consumer interest govt. provides specified quantity of goods to the consumer at the prices which are lower than market prices.
- *Procurement Prices :-* Govt. every year maintains a particular stock of a product. This stock is used at time to time of emergency and shortage for the purchase of the desired agricultural product govt. announces procurement prices. These prices are generally reasonable.

Changing in Demand and Supply

Some times govt. producers some portion of the product it reduces the supply in the open market then price level rises. Govt. allows the sellers to sell their product at market price. Sometime govt. exports the surplus product. It reduces the supply, price level rises. The farmer sell product at reasonable price.

Improvement in Facilities

The govt. can influence the price of the product by providing the facility of where house and roads and markets. Due to these facilities farmer can sell his product at reasonable price. Most of the different countries have also revised its agricultural price policy keeping in view the problems of the farmers and consumers. The price support policy is implemented through various departments. Food, Cotton and Rice export corporation. Agricultural marketing and storage limited, and ghee corporation are playing very effective role in this regard. Wheat, rice, cotton, sugarcane, sunflower etc. Prices are set through the price support policy.

PRICE POLICY AND AGRICULTURAL MARKETING

The Government of India intervenes in the agricultural markets to achieve certain developmental objectives. The overarching reasons for effective government interventions are stated to be food security and price stability. The government intervenes in domestic market in various forms such as food grain procurement and distribution, price support, input subsidies and marketing legislations. The objectives and forms of intervention have undergone substantial changes over time. The interventions attempted to bring in regulation of various agricultural activities to protect the interests of producers and consumers. But, such regulations did not foster a competitive environment for fair play of market forces.

PROCUREMENT AND DISTRIBUTION

The Government of India's food grain policy aims at achieving reasonable price support and procurement system to increase farm income and making available food grains to consumers at reasonable price through distribution of subsidised food grains and price stabilisations/ buffer stock operations. The Food Corporation of India (FCI) is entrusted with implementation of food grains policy particularly for rice and wheat. FCI or the designated agency of state government procures paddy and wheat from the farmers at minimum support price (MSP). Additionally, FCI procures rice through a levy system from rice mills. Depending on the state, rice mills are required to deliver to the FCI from 10 to 75 per cent of their milled rice at the prescribed levy price. Wheat and paddy/ rice procured thus are used to meet the demand for public distribution system, buffer stocks and other welfare measures. FCI's operations are intended to build buffer stocks to meet any exigency, open market sales to stabilise the domestic price and to meet the food security requirements. In general, the official procurement operations are carried out through regulated markets set up under APMC Act. A network of regulated markets was created to promote organised marketing of agriculture produce. Except Kerala and Manipur, all the other states had enacted State level APMC Acts. In 2005, there were about 7,557 regulated markets spread across various states in India. The geographical distribution of markets was skewed towards large states: larger the size of area, more the number of markets. States like Andhra Pradesh, Bihar, Maharashtra, Madhya Pradesh, Uttar Pradesh and West Bengal had share of more than 50 per cent of total number of markets. The regulated markets handled about 20 per cent of total marketed surplus.

During initial periods of operations, the regulated markets helped to mitigate the difficulties faced by the agricultural producers in disposing their produce. Despite several drawbacks in the functioning of regulated marketing system, they helped to provide access to the markets and increase income of the farmers. But over a period of time, these regulated markets failed to serve the interests of the farmers in a reasonable manner. Some of the rigidities incorporated in the Act bred inefficiency in the system. There are instances that agricultural produces are marketed bypassing the regulated market yards.

Several studies and Committees pointed out restrictive provisions of Act and their impact on the efficient functioning of the market. Some of the obstacles posed by APMC Acts are summarised as follows:

- The restrictive legal provisions like delineation of "market area" do not promote a competitive market structure. Farmers do not have options to sell his produce at any other place/agency than regulated market. This has not facilitated the direct supply to the processing and consuming industries and has hampered the development of retail supply chain.
- The powers vested with market committees to issue and suspend/ cancel the licenses granted to traders have resulted in ethical practices of arbitrage and favouritism. The basic functions of regulations, correct weighing and proper sale had not been given much importance.
- The licensed traders acquired monopoly status by forming collusion among them. The monopoly status in marketing and handling has added to marketing costs detrimental to producers and consumers. The new entrants find very difficult to operate under such collusive environment, thus stifling the competition. The monopolistic circumstances also do not allow use of latest technologies in handling, grading, packaging and transportation.
- State Agricultural Marketing Board undertakes market infrastructure development. Market committees contribute some percentage of their income to the Board in this regard. But, it is reported that funds from the Boards are siphoned off to Public Ledger Account of State governments, thereby limiting the development of market infrastructure.
- The regulated markets are mostly located in towns and remained out of reach of farmers living in far-flung areas. Though, the density of regulated markets varies across the states, but on average a market serves 459 sq. Km in the country, which is quite high.
- Market infrastructure is important for performing various marketing operations efficiently. The facilities available in the market are not adequate. For instance, only one fourth of markets have common drying yards. Cold storage facilities exist only in 9 per cent of markets and grading facilities in less than one third of the market.
- Farmers are represented in the Market Committees of regulated markets. But their voice is rarely effective and lack control over certain marketing functions.

FOOD SUBSIDY

The economic cost of the process of food procurement and distribution includes three components, viz., price paid to the farmers, procurement operations and the cost of distribution. The difference between economic cost of foodgrains and the issue price of FCI is equivalent food subsidy. Food subsidy provided to FCI and decentralised state-level procurement operations increased from ₹ 92 billion in 1999-2000 to ₹ 241.2 billion in 2002-03 and then up to ₹ 436.7 billion in 2008-09.

REFORM OF APMC ACT

Observing the inefficiency caused by licensing/registration, market controls and other interference introduced by APMC Acts, it was strongly felt that an alternative marketing system needs to be introduced. The government should facilitate smooth operations of the markets and should not control over it. Further, greater participation of private sector should be encouraged to make investments required for the development of marketing infrastructure and other supporting services.

For the first time a National Policy on Agriculture was introduced in July 2000. The National Policy aims to attain agricultural growth rate of over 4 per cent per annum in the next two decades. The National policy also underlines bringing about domestic market reforms to create favourable economic environment and to remove the distortion. Subsequently, in December 2000, an Expert Committee was constituted to suggest policy recommendations to strengthen the agricultural marketing. The Committee submitted its report in 2001. To examine findings and recommendations of the Expert Committee and to suggest measures to implement them, the Ministry of Agriculture, Government of India constituted an Inter-Ministerial Task Force on July 2001. The important recommendations/measures proposed by Task Force are as follows.

- An alternative marketing systems should be developed to promote competition by amending State APMC Acts; Central assistance should be provided for development and strengthening of general and commodity specific agricultural produce markets
- Progressive dismantling of controls and regulations under the Essential Commodities Act
- The institutional credit for marketing of crops (pledge financing) should be increased; negotiability status should be given to warehousing receipts of agricultural commodities.
- Allowing of futures trading in all agricultural commodities to improve price risk management and facilitate price discovery by amending the Forward Contracts (Regulation) Act, 1952

These recommendations were discussed with State governments. Most of the states expressed the view that reforms in the agricultural markets were necessary to move away from the regime of controls to one of regulation and competition. Accordingly, Ministry of Agriculture, Government of India in consultation with state governments formulated a Model Act, called "The State Agricultural Produce Marketing (Development and Regulation) Act, 2003. The important features of the Model Act are given below:

- Permission for establishment of Private Markets/Yards, Direct Purchase Centres, Consumer/Farmers Markets for direct sale and promotion of Public Private Partnership in the management and development of agricultural markets in the country.
- Provisions for separate constitution of Special Markets for Commodities like Onions, Fruits, vegetables, Flowers etc.
- A separate Chapter has been included in the legislation to regulate and promote contract-farming arrangements in the country.
- It provides for prohibition of commission agency in any transaction of agricultural commodities with the producers.
- It redefines the role of present Agricultural Produce Market Committee to promote alternative marketing system, contract farming, direct marketing and farmers/consumers markets.
- It also redefines the role of State Agricultural Marketing Boards to promote standardisation, grading, quality certification, market led extension and training of farmers and market functionaries in marketing related areas.
- Provision made for resolving of disputes, if any arising between private market/ consumer market and Market Committee.
- Provision has also been made in the Act for constitution of State Agricultural Produce Marketing Standards Bureau for promotion of Grading, Standardisation and Quality Certification of agricultural produce. This would facilitate pledge financing, E-trading, direct purchasing, export, forward/future trading and introduction of negotiable warehousing receipt system in respect of agricultural commodities.

All the State governments have been asked to bring changes in the APMC Acts on the lines of Model Act. The progress of the reforms in APMC Acts. Several state governments have taken initiatives to adopt the model Act. Karnataka permitted the National Dairy Development Board to set up a fruit and vegetable wholesale market. State governments such as Punjab, Haryana, Madhya Pradesh and Tamil Nadu allowed contract farming arrangements to enable the farmers to sell directly to private buyers.

Besides, APMC Acts there are several other legal instruments used by both central and state governments to regulate the functioning of agricultural markets. Amongst various Acts the most pervasive one has been the Essential Commodities Act, 1955. Most of restrictions related to movement, storage, processing and stock limits are contained in this Act. The operation of ECA has created obstacles in the free flow of commodities from surplus to deficit region. This has widened the price wedge between the different parts of the country and increased the cost of marketing. ECA has prevented large scale participation of private traders in various marketing activities. In fact, private investment in large-scale storage and marketing has been non-existent due to restrictive provisions of this Act and Control Orders issued thereof. Presently, there are about 15 essential commodities are covered under this Act.

The Food Safety and Standards Act, 2006 consolidates different Acts relating to food. These Acts include: The Prevention of Food Adulteration Act, 1954; The Fruits Products Order, 1955; The Meat Food Products Order, 1973; The Vegetable Oil Products (Control) Order, 1947; The Edible Oils Packaging (Regulation) Order, 1998; The Solvent Extracted Oil, De oiled Meal and Edible Flour (Control) Order, 1967; The Milk and Milk Products Order, 1992 and any other order issued under Essential Commodities Act, 1955 relating to food.

The Food Safety and Standard Authority of India (FSSAI) has been established under Food Safety and Standards Act, 2006. It consolidates various acts and orders that have hitherto handled food related issues in various Ministries and Departments. FSSAI has been created for laying down science based standards for articles of food and to regulate their manufacture, storage, distribution, sale and import to ensure availability of safe and wholesome food for human consumption.

In recent times, there is huge enthusiasm generated regarding forward trading of agricultural commodities and is expected to provide price stability in the domestic market. Forward Markets (Regulation) Act, 1952 regulates the commodity futures markets in India. The Forward Markets Commission (FMC) performs the functions of advisory, monitoring, supervision and regulation in futures and forward trading.5 Futures' trading is conducted in exchanges owned by the private associations registered under the Act. These exchanges operate independently under the guidelines of their bylaws approved by the FMC. Futures' trading in agricultural commodities is a recent phenomenon. Government has permitted futures trading in 54 agricultural commodities with effect from April 2003. Futures are traded in 24 commodity exchanges of which 3 are national exchanges and 21 are regional exchanges. Agricultural commodities constitute over 60 per cent of total volume of trade in 2005-06. The major commodities traded in futures market in terms of decreasing order of value are guar seed, chana (gram), urad (black gram), soy oil, tur (red gram), menth oil and guar gum. The participation of traders and farmers in the futures trading is very much limited due to uncertainty in the policies of government of India. The government uses futures trading as one of the instruments to contain raise in prices of agricultural commodities. In February 2007, government banned futures trading in rice and wheat. Further, in May 2008, four more commodities such as potato, gram (*Chana*), soya oil and rubber have been added to the list of banned commodities.

Commodity trade in futures markets in 2009 included various agricultural commodities, bullion, crude oil, energy and metal products. Some new commodities were included including almond, imported thermal coal, platinum and carbon credits. The average daily value of trade in commodity exchanges increased from ₹ 164 billion in 2008 to ₹ 232 billion in 2009.

PRICE SUPPORT

Government's price support policy provides guarantee against sharp fall in commodity prices and helps ensure reasonable income to farmers. Presently, government sets minimum support price for 25 commodities. The Commission for Agricultural Costs and Prices (CACP) recommends the MSP for these commodities. The MSP of various commodities has increased rapidly during recent years. However, the operation of MSP system has been controversial as it draws criticism from different quarters of policy makers and academics for its basis of calculation and its relevance in the present context. The minimum support prices, especially for foodgrains, are being effectively implemented only in a few surplus states like Punjab, Haryana, Uttar Pradesh and Andhra Pradesh and only a small segment of farmers in the country are benefited. The government need not bear the commodity price risks once the income safety nets for the poor have effectively been put in place.

The average annual income transfer per household from the government wheat price support programme at C2 Cost was found to be ₹.9,980 in Punjab and ₹ 5,790 in Haryana. In case of rice it was ₹ 3041 in Punjab and ₹ 164 in Andhra Pradesh. Further, the income transfers mainly accrue to the large farmers. Whether for rice or wheat, the average income transfer to large farmers is over 10 times greater than those received by marginal farmers.

REFORMS IN AGRICULTURE MARKET

To protect the farmers' interests and to ensure adequate availability of food grains for official procurement from the unfair practices by private grain traders, a large number of restrictions were imposed by the central and state government.

However, excessive regulations of domestic marketing had resulted in increased marketing costs, risks and uncertainty, which impacted the performance of agriculture sector. Excessive regulations dampen growth through suppressing competition in the market. Further, India has achieved food security at the national level through its interventionist policies in input and output markets. However, it is not sustainable to continue with controlled markets for long time. To achieve higher growth and move the sector in long term sustainable growth trajectory, market reforms in agriculture need to be strengthened. The government of India initiated domestic market reforms in agriculture since, 2000. These included improving the performance of commodity markets, reforms in APMC Act, reforms in price policy, rationalisation of input subsidies, increasing public investments, operation of futures trading and encouraging participation of private sector.

Necessary changes have been effected to progressively dismantling controls and regulations under Essential Commodities Act. In February 2002, Government of India brought out Removal of (Licensing requirements, stock limits and movement restrictions) on Specified Foodstuffs Order. This order removed all restrictions and licensing of dealers on purchase, stocking, transport of the specified commodities viz., wheat, paddy/rice, coarse grains, sugar, edible oil seeds and edible oils. It also specifies that issuing of any order by the states/ UTs under ECA for regulating licensing, storage, transport and distribution of any of the specified commodities will require the prior concurrence of the Central Government.

The trade policy reforms progressed during 1990s to facilitate greater integration of agriculture sector with global market. Quantitative restrictions on import of agriculture commodities were removed in April 2001. The list of commodities which were earlier canalised through State Trading Enterprises was trimmed down. The average import tariffs were also reduced considerably over time. Agricultural export policies were liberalised in 1994. It included relaxation in export quotas, abolition of minimum export prices and increased availability of credit. To boost agricultural exports, "Vishesh Krishi Upaj Yojana" (Special Agricultural Produce Scheme) was introduced in EXIM policy 2002-07 with special incentives. However, the Government often tampers with export policies to make available commodity supplies to stabilise the price level in the domestic market.

PRICE POLICY IN AGRICULTURAL PRODUCTS

The government has formulated a price policy for agricultural produce that aims at securing remunerative prices to farmers to encourage them to invest more in agricultural production. Keeping this in mind, the government announces minimum support prices for major agricultural products every year. These prices are fixed after taking into account the recommendations of the Commission for Agricultural Costs and Prices (CACP).

The Commission of Agricultural Costs and Prices while recommending prices takes into account important factors, such as:

- Cost of production
- Changes in input prices
- Input/Output Price Parity
- Trends in market prices
- Inter-crop Price Parity
- Demand and supply situation
- Effect on Industrial Cost Structure

- Effect on general price level
- Effect on cost of living
- International market price situation
- Parity between prices paid and prices received by farmers (Terms of Trade)

PRICING PROJECT COSTS AND BENEFITS

Once costs and benefits have been identified, if they are to be compared they must be valued. Since, the only practical way to compare differing goods and services directly is to give each a money value, we must find the proper prices for the costs and benefits in our analysis.

PRICES REFLECT VALUE

Underlying all financial and economic analysis is an assumption that prices reflect value - or can be adjusted to do so. Before proceeding, however, it is necessary to define two economic concepts crucial to project analysis: marginal value product and opportunity cost.

Consider a Filipino farmer who applies nitrogenous fertilizer to his rice. In the 1979-80 season this fertilizer cost him P3.98 per kilogram of elemental nitrogen (N), and he received P1.050 for every kilogram of paddy rice he sold. (The symbol for Philippine pesos is P.) Table 4.1 shows the responsiveness of his rice to fertilizer. At low levels of application, fertilizer has a great effect on rice yield. Increasing the application from no fertilizer to 10 kilograms of elemental nitrogen increased the farmer's yield from 3,442 kilograms to 3,723 kilograms per hectare and increased the value of his output by P295, from P3,614 to P3,909.

Table. 4.1: Crop Response to Nitrogen Fertilizer in the Philippines.

	Paddy rice			Shelled maize		
Nitrogen (kgs/ha)	Yield (kgs/ha)	Value	MVP	Yield (kgs/ha)	Value	MVP
0	3,442	3,614		2,600	2,688	
10	3,723	3,909	29.50	2,830	2,926	23.80
20	3,971	4,170	26.10	3,040	3,143	21.70
30	4,187	4,396	22.60	3,230	3,340	19.70
40	4,370	4,588	19.20	3,400	3,516	17.60
50	4,520	4,746	15.80	3,550	3,671	15.50
60	4,637	4,869	12.30	3,680	3,805	13.40
70	4,721	4,957	8.80	3,790	3,919	11.40
80	4,772	5,011	5.40	3,880	4,012	9.30
90	4,791	5,031	2.00	3,950	4,084	7.20
100	4,777	5,016	-1.50	4,000	4,136	5.20
110	4,030	4,167	3.10	120	4,040	4,177
1.00	130	4,030	4,167	-1.00		

Thus, for every additional kilogram of elemental nitrogen the farmer applied at this level, he received P29.50 in return [(3,909 - 3,614) _ 10 = 29.50]. The extra revenue from increasing the quantity of an input used, all other quantities remaining constant, is the marginal value product of the input. In this case, then, the marginal value product of a kilogram of fertilizer is P29.50.

If the farmer could buy fertilizer for P3.98 a kilogram and use it to increase output by PP9.50, it obviously would have paid him to apply more. But as the intensity of application increases, each additional kilogram of fertilizer has less and less effect on production. If the farmer had increased his application from 80 to 90 kilograms per hectare, he would have increased the value of his production by only P20, from P5,011 to P5,031, and the marginal value product of a kilogram of fertilizer would have fallen to only P2.00 [(5,031 - 5,011) - 10 = 2.00]. Since, he would have had to pay P3.98 per kilogram, it clearly would not have been worthwhile to apply fertilizer at this rate. In fact, it would only have paid the farmer to apply fertilizer up to the rate at which the marginal value product just equaled the price. For this Filipino farmer, it would have paid him to apply approximately 80 kilograms of nitrogen: between 70 and 80 kilograms the marginal value product of each additional kilogram was some P5.40, whereas between 80 and 90 kilograms it fell to P2.00. Thus, the farmer would have expanded his fertilizer use until he reduced the marginal value product of the fertilizer to its market price, and the market price, therefore, is an estimate of the marginal value product of the fertilizer.

The optimal amount of fertilizer to use will change, of course, when the price of fertilizer changes relative to the price of rice. If the relative price of fertilizer were to rise, the farmer would respond by reducing the amount of fertilizer he applies, increasing the marginal value product of the fertilizer (but reducing the total amount and value of production) until the marginal value product of the fertilizer again just equals its price. Suppose fertilizer were to double in price to P8.00 per kilogram of elemental nitrogen, and rice prices remained unchanged. Then, the farmer should reduce the amount of fertilizer applied to a hectare from 80 kilograms to 70 kilograms, since, between 60 and 70 kilograms the marginal value product was some P8.80 but between 70 and 80 kilograms it was only some P5.40.

In practice, because of risk and limited resources, the farmer would probably not have applied the amounts indicated here. We may consider that the farmer reduces his expected return by some "risk discount." Even so, the principle we are illustrating remains the same: the farmer equates the expected marginal value product less some risk discount to the price of fertilizer.

If this farmer also grew maize, for which in 1979-80 he would have received P1.034 per kilogram of shelled grain, it would have paid him (in the absence of risk) to apply some 100 kilograms of elemental nitrogen to each hectare, because between 90 and 100 kilograms the marginal value product of a kilogram of nitrogen applied to maize was P5.20, whereas between 100 and 110 kilograms the marginal value product fell to P3.10, below the price of fertilizer.

Now, suppose the farmer had limited resources and could not obtain sufficient credit to increase his fertilizer application on both rice and maize to where the marginal value product equaled the price. Suppose the farmer had only 2 hectares, 1 planted in rice and 1 in maize, and resources sufficient to purchase just 80 kilograms of nitrogen. How should he have used it? Should he have put it all on rice and none on maize? If he did, he would have applied fertilizer to his rice at the level where the marginal value product was just about equal to its market price. But suppose he had shifted some fertilizer, instead, to maize. If he had shifted 10 kilograms, he would have reduced the value of his rice production by P54- from P5,011 to P4,957, or by P5.40 for each kilogram shifted-but he could have obtained some P238 for the 10 kilograms applied to maize, since, the marginal value product between 0 and 10 kilograms was some P23.80 per kilogram. In other words, at these levels each kilogram of nitrogen shifted would reduce the rice value by P5.40 but increase the value of maize output by some P23.80. In the language of economics, the opportunity cost of fertilizer shifted from rice to maize was P5.40. Opportunity cost, thus, is the benefit forgone by using a scarce resource for one purpose-in this case applying fertilizer to maize-instead of for its best alternative use-in this case using the fertilizer to produce rice. Said another way, the opportunity cost is the return a resource can bring in its next best alternative use. What would be the opportunity cost if the farmer were to move a kilogram of fertilizer in the other direction, back from maize to rice? He would have given up P23.80 to gain only P5.40-not a very attractive proposition-and the opportunity cost, obviously, would be some P23.80.

Given his limited resources, it would pay the farmer to shift fertilizer from rice to maize until the marginal value product of fertilizer applied to both crops is the same. In the case of the Filipino farmer who could buy only 80 kilograms of fertilizer, if on the one hand he were to move 40 kilograms to maize, reducing his application on rice from 80 kilograms to 40 kilograms, he would have increased the marginal value product of the fertilizer on his rice to some P15. On the other hand, the 40 kilograms shifted away from rice and put on maize would have decreased the marginal value product of nitrogen applied to maize also to about P15. At these levels, there would be no advantage in shifting fertilizer between the two crops-the opportunity cost of shifting more fertilizer from rice to maize would be about P15, but the gain would also be only about P15-and the farmer would have reached the optimal level of application to both crops.

Note, however, that if the farmer could somehow have bought as much fertilizer as he wanted at the market price of P3.98 per kilogram-perhaps through a credit programme-then the market price of fertilizer would have become its opportunity cost, and (in the absence of a risk discount) he should have increased his application to 80 kilograms on rice and 100 kilograms on maize. From a single farmer to the economy as a whole, the same principles apply. In a "perfect" market-one that is highly competitive, with many buyers and sellers, all of whom have perfect knowledge about the market-every economic commodity would be priced at its marginal value product, since, every farmer will have expanded his fertilizer use to where its marginal value product equals its price, and the same will have happened for every other item in the economy. That is, the price of every good and service would exactly equal the value that the last unit utilized contributes to production, or the value in use of the item for consumption would exactly balance the value it could contribute to additional production. If a unit of goods or services could produce more or bring greater satisfaction in some activity other than its present use, someone would have been willing to bid up its price, and it would have been attracted to the new use. When this price system is in "equilibrium," the marginal value product, the opportunity cost, and the price will all be equal. Resources will then have been allocated through the price mechanism so that the last unit of every good and service in the economy is in its most productive use or best consumption use. No transfer of resources could result in greater output or more satisfaction.

Without moving further into price theory, we can consider some direct implications for agricultural projects of the assumption that prices reflect value.

First, as everyone knows, markets are not perfect and are never in complete equilibrium. Hence, prices may reflect values only imperfectly. Even so, there is a great deal of truth in this price theory based on the model of perfect markets. In general, the best approximation of the "true value" of a good or service that is fairly widely bought and sold is its market price. Somebody in the economy is willing to pay this price.

One can presume that this buyer will use the item to increase output by at least as much as its price, or that he is willing to exchange something of value equal to the price to gain the satisfaction of consuming the item. Hence, the market price of an item is normally the best estimate of its marginal value product and of its opportunity cost, and most often it will be the best price to use in valuing either a cost or a benefit. In financial analysis, as we have noted, the market price is always used. But in economic analysis some other price-a "shadow price"-may be a better indicator of the value of a good or service; that is, a better estimate of its true opportunity cost to the economy. When prices other than market prices are used in economic analysis, however, the burden of proof is on the analyst.

FINDING MARKET PRICES

Project analyses characteristically are built first by identifying the technical inputs and outputs for a proposed investment, then by valuing the inputs and outputs at market prices to construct the financial accounts, and finally by adjusting the financial prices so they better reflect economic values. Thus, the first step in valuing costs and benefits is finding the market prices for the inputs and outputs, often a difficult task for the economist.

To find prices, the analyst must go into the market. He must inquire about actual prices in recent transactions and consult many sources-farmers, small merchants, importers and exporters, extension officers, technical service personnel, government market specialists and statisticians, and published or privately held statistics about prices for both national and international markets. From these sources the analyst must come up with a figure that adequately reflects the going price for each input or output in the project.

POINT OF FIRST SALE AND FARM-GATE PRICE

In project analysis, a good rule for determining a market price for agricultural commodities produced in the project is to seek the price at the "point of first sale." If the point of first sale is in a relatively competitive market, then the price at which the commodity is sold in this market is probably a relatively good estimate of its value in economic as well as financial terms. If the market is not reasonably competitive, in economic analysis the financial price may have to be adjusted better to reflect the opportunity cost or value in use of the commodity.

For many agricultural projects in which the objective is increased production of a commodity, the best point of first sale to use is generally the boundary of the farm. We are after what the farmer receives when he sells his product-the "farm-gate" price. The increased value added of the product as it is processed and delivered to a market arises as a payment for marketing services. This value added is not properly attributed to the investment to produce the commodity. Rather, it arises from the labour and capital engaged in the marketing service. Usually the price at point of first sale can be accepted as the farm-gate price; even if this point is in a nearby village market, the farmer sells his output there and thus earns for himself any fee that might be involved in transporting the commodity from the farm to the point of first sale. But if any new equipment is necessary to enable the farmer to do this-say, a new bullock cart or a new truck-then that new equipment must be shown as a cost incurred to realize the marketing benefit in the project.

In projects producing commodities for well-organized markets, the farm-gate price may not be too difficult to determine. This would be true for most food grains traded domestically in substantial quantities. One may think of wheat in most countries of the Middle East and South Asia, of rice in South and Southeast Asia, and of maize in much of Latin America. It would also be true of farm products for which the processor is generally the first buyer (such as fresh fruit bunches for palm oil in Malaysia or milk in Jamaica), where the price quoted to the farmer is the price on his farm, and the firm responsible for the marketing comes to the farm to pick up the product. In many cases, however, the prices in a reasonably competitive market or in the price records kept by the government statistical service will include services not properly attributable to the investment in the project itself. This may happen, for instance, when the only price series available for a product records the prices at which it has been sold in a central market-such as the price for eggs in Madras, for melons in Tehran, or for vegetables in Bogota. In that case, the project analyst will have to dig deeper to find out how to value the marketing services. Then he can adjust the central market price to reduce it to the farm-gate price. The farm-gate price is generally the best price at which to value home-consumed production. In some cases it may be extremely difficult to determine just what a realistic farm-gate price is for a crop produced primarily for home consumption because so little of the crop appears on markets. This is the case, for example, for manioc and cocoyam in Africa. On the one hand, some argue that the true value of the crop is overstated if the market price is used as a basis for valuation because such a small proportion of the product is actually sold. On the other hand, the same crop in different situations may not be so difficult to value. Manioc is sold extensively in Nigeria to make gari flour, and it is commonly traded in local markets in tropical Latin America and the Caribbean.

The farm-gate price may be a poor indicator of the true opportunity cost we want to use in economic analysis. In Ghana the Marketing Board takes some proportion of the cocoa price as a tax for development purposes. In Thailand, a rice "premium" - that is, a tax on rice exports-effectively keeps the domestic price well below what the international market would pay. In these cases, when the commodity is traded its economic value would have to be considered higher than the actual farm-gate price, and this price distortion will have to be corrected in the economic analysis. In other cases, just the opposite happens. In Mexico the price of maize is maintained at a high level to transfer income to ejidatarios, the small farmers. In Malaysia, the price of rice is supported above world market levels to encourage local production and to reduce imports. In these cases, part of the price does not really reflect the economic value of the product-its cost if it could be imported-but rather an indirect income transfer to small farmers. Again, this price distortion will have to be corrected in the economic analysis.

PRICING INTERMEDIATE GOODS

By emphasizing the point of first sale as a starting point for valuing the output of our projects, we are also implying that imputed prices should be avoided for intermediate goods in our analysis. An intermediate good is an item produced primarily as an input in the production of another good. If an intermediate good is not freely traded in a competitive market, we cannot expect to obtain a price established by a range of competitive transactions. Fodder produced on a farm and then fed to the dairy animals on the farm is an example of such an intermediate product. If increased fodder production is an element in the proposed agricultural project, the analyst would avoid valuing it. Instead, the analyst would treat the whole farm as a unit and value the milk produced at its point of first sale or value the calves sold as feeder cattle. Treatment of intermediate products will vary from project to project depending on the particular marketing structures. In some countries it would hardly make sense in an egg production project to value the pullets produced in a pullet production enterprise and then "sell" these pullets to the egg production enterprise on the same farm. But in other countries there might be an active market in pullets, which would mean that we could expect to find a reasonably competitive price to use in the economic analysis. To avoid most of the problems that might be introduced by trying to impute values for intermediate products, the financial accounts in agricultural projects are based on budgets for the whole farm instead of on budgets for individual activities on the farm; that is, on the budget for the egg farm as a whole rather than on the budget for a pullet production activity.

A frequently encountered intermediate good in agricultural projects is irrigation water. The "product" of an irrigation system-water-is, of course, really intended to produce agricultural commodities. The price farmers are charged for the water is generally determined administratively, not by any play of competitive market forces. If the analyst were to try to separate the irrigation system from the production it makes possible, he would be faced with a nearly impossible task of determining the value of irrigation water. Hence, it is not surprising that the economic analyses of most irrigation projects take as the basis for the benefit stream the value of the agricultural products that are offered in a relatively free market at the point of first sale.

OTHER PROBLEMS IN FINDING MARKET PRICES

Considerable confusion often arises in determining the values for two important inputs in agricultural projects, land and labour. This happens primarily when the analysis moves from the financial project accounts to the economic analysis. In the accounts prepared for the financial analysis, the treatment of prices for land and labour is quite straightforward: the price used is the price actually paid. Thus, if the farmers in a settlement project are expected to pay the project authority a price for the land they acquire, perhaps through a series of installments, then the actual price in the year it is paid is entered in the project accounts. In the financial analysis, we do not question whether this is a "good" price in economic terms. Similarly, if land must be bought for the right-of-way for canals in an irrigation project, the actual price to be paid is entered in the project accounts in the financial analysis. Or, if the project includes tenant farmers who will receive help in increasing wheat production, then in the financial accounts for these tenant farmers the analyst will enter the rent paid each year at the amount actually paid, or at the farm-gate value of the wheat delivered to the landowner if the tenants pay rent in kind.

If farm accounts are laid out on a with - and - without basis following the format, in those instances where the project involves only changing the cropping pattern (say, a shift from pasture to irrigated sorghum), the cost of the land (in this instance an opportunity cost) need not be separately entered because of the form of the account. When the net benefit without the project is subtracted from the net benefit with the project, the contribution of the land to the old cropping pattern is also subtracted and only the incremental value remains.

In valuing labour for the financial analysis accounts, again, the problems arise when the financial accounts are adjusted to reflect economic values. For financial analysis, the analyst enters the amounts actually paid to hired labour, either in wages or in kind, in the farm budgets or project accounts. Family labour is treated differently. It is not entered as a cost; instead, the "wages" for the family become a part of the net benefit. Thus, if our project increases the net benefit, it also in effect increases the family's income or "wages" for its labour. The account will automatically value the family labour at its opportunity cost, and the incremental net benefit will reflect any increased return the family may receive for its labour.

Prices for agricultural commodities generally are subject to substantial seasonal fluctuation. If this is the case, some decision must be made about the point in the seasonal cycle at which to choose the price to be used for the analysis. A good starting point is the farm-gate price at the peak of the harvest season. This is probably close to the lowest price in the cycle. The line of reasoning here is that as prices rise during the cycle at least some part of that rise is a result not of the production activities of the farmer but of the marketing services embodied in storing the crop until consumers want it. But, markets being what they are, there may be an element of imperfection in the harvest price level. Market channels may become so glutted that merchants try actively to discourage farmers from immediately bringing their crop to the market by offering a price that even the merchants themselves would admit is too low. Even so, the need to sell immediately to meet debt obligations may force farmers to offer their crops despite these artificially low, penalty prices. In some cases, therefore, a price higher than the farm-gate price in the harvest season may be selected. But there is an obligation here to justify the price chosen as more valid than the lowest seasonal price. One way to resolve this problem may be to include an element of credit in the project design. This would permit farmers to withhold their product from the market until prices have had a chance to rise from their seasonal lows but at the same time to have enough money to meet their cash obligations and family living expenses. The credit element may also include credit for building on-farm storage so that farmers will have a safe place to store their production until they decide to market it at a better price.

Prices vary among grades of product, of course, and picking the proper price for project analysis may involve making some decisions about quality of the product. In general, it can be assumed that farmers will produce in the future much the same quality as they have in the past and will market their product ungraded. In many agricultural projects, however, one objective is to upgrade the quality of production as well as to increase the total output. Small dairy farmers, for instance, may be able with the help of the project investment to meet the sanitation standards of the fluid milk market and to command a higher price; or reduced time for delivery may hold down sucrose inversion in sugarcane; or better pruning will increase the average size of the oranges Moroccan farmers can offer European buyers. In such cases, the proper price to select is the average price expected for the quality to be produced.

A special problem occurs in pricing housing. If project investment includes housing construction, as would be the case for a settlement project, then one benefit arising from the investment is the rental value of the house. Since, the rental value will usually be an imputed value rather than a real market price, care must be exercised in determining it. No more should be allowed for the rental value than would normally be paid by a prospective tenant family. Nor should more rental value be allowed than the family would be expected to pay for a comparable house in the vicinity or in a similar area elsewhere (if the new settlement is in a distant locale). In particular, the temptation should be avoided to take as a rental value some arbitrary proportion of the housing cost. Otherwise, overly elaborated housing construction might be justified simply by assigning it an unrealistically high imputed value.

PROJECT BOUNDARY PRICE

Prices used in analysing agricultural projects are not necessarily farm-gate prices. The concept of a farm-gate price may be expanded to a "project boundary" price if a project has a marketing component or if it is a purely marketing project. Many projects have a marketing component, perhaps because there is no competitive channel reaching down to the farm-gate level for the unprocessed product. Of concern in these projects are both the farm-gate price (on which to base the estimates of the net benefit to the farmer) and the price at which the processed product is sold in the market (after being handled in the facilities financed by the project). Such a case is found in the Rahad project in the Sudan. There the Roseires dam on the Blue Nile will provide irrigation water for the production of cotton, which will be ginned in new facilities financed by the project. The analyst, of course, is interested in the price of cotton paid to the farmers so that their incomes can be estimated. But, since, this price is set administratively, it could not be used directly in the economic analysis of the project. The analyst is also interested in the price of ginned cotton because that is the first product the project will actually sell in a reasonably competitive market. In this case, the point of first sale is f.o.b. (free on board) Port Sudan, and the price there becomes the basis for the benefit stream.

PREDICTING FUTURE PRICES

Since, project analysis is about judging future returns from future investment, as analysts we are immediately involved in judging just what future prices may be. This is a matter of judgement, not mechanics. No esoteric mathematical model exists to come to the aid of the project analyst; like everyone else he must take into consideration all the facts he can find, seek judgements from those he respects, and then come to a conclusion himself. It tends to be a rather unsettling process. The only consolation is that careful, considered judgement about the course of future prices is better than giving the matter no thought at all and wasting scarce resources on incompletely planned projects.

We have been discussing how to find market prices, and it is from these current prices that we begin. The best initial guess about future prices is that they will retain the present relationships, or perhaps the average relationship they have borne to each other over the past few years. We must consider, however, whether these average relationships will change in the future and how we will deal with a general increase in the level of prices owing to inflation.

CHANGES IN RELATIVE PRICES

We may first raise the question of whether relative prices will change. Will some inputs become more expensive over time in relation to other commodities? Will some prices fall relatively as supplies become more plentiful? Not easy questions to deal with, but some approaches to answers can be made. In financial analysis, of course, a change in a relative price means a change in the market price structure that producers face either for inputs or for outputs.

A change in a relative price, then, is reflected directly in the project's financial accounts. A rise in the relative price of fertilizer reduces the incremental net benefit-the amount the farm family has to live on. It is thus clearly a cost in the farm account. The same line of reasoning can be applied in the financial analysis for any other group participating in the project.

A change in the relative price of an item implies a change in its marginal productivity-that is, a change in its marginal value product-or a change in the satisfaction it contributes when it is consumed. In economic analysis, where maximizing national income is the objective, a change in the relative price of an input implies a change in the amount that must be forgone by using the item in the project instead of elsewhere in the economy; it is therefore a change in the contribution the output of the project makes to the national income. Thus, changes in relative prices have a real effect on the project objective and must be reflected in project accounts in the years when such changes are expected.

There are several kinds of commodities subject to future changes in relative prices. Most agricultural project analysts would probably agree that the relative price of energy-intensive agricultural inputs is likely to continue to rise over the next several years, just as it has done over the past few years. Thus, on the input side the project accounts might show an annual increase, at least for the first decade or so, in the cost of fuel for tractors, for transporting the harvested crop, for drying grain, and for such petroleum-based inputs as fertilizers and chemical pesticides. On the output side, there may be some commodities that will probably continue to be in short supply and whose prices will rise as incomes increase - one might think of mutton from fat-tailed sheep in Iran, or, for that matter, of most meat products worldwide. How much will prices increase relative to those of other products? Certainly a difficult question, but one the project analyst must confront. For a range of products-from industrial crops such as fibres or oilseeds to food grains and vegetables-judgements will have to be made on the best possible basis.

In some countries, relative wages of rural labour may rise as economic development proceeds during the life of a project. This will have implications not only for the prices assumed for hired labour, but also for the incentive effect exerted by a given change in net benefit and for the technology assumed as a basis for projections in the farm budgets and project accounts.

INFLATION

In the past few years, virtually every country has experienced inflation, and the only realistic assessment is that this will continue. No project analyst can escape deciding how to deal with inflation in his analysis.

The approach most often taken is to work the project analysis in constant prices. That is, the analyst assumes that the current price level (or some future price level - say, for the first year of project implementation) will continue to apply. It is assumed that inflation will affect most prices to the same extent so that prices retain their same general relations. The analyst then need only adjust future price estimates for anticipated relative changes, not for any change in the general price level. By comparing these estimates of costs and benefits with the constant prices, he is able to judge the effects of the project on the incomes of participants and its income-generating potential for the society as a whole. Although the absolute (or money) values of the costs and benefits in both the financial and the economic analyses will be incorrect, the general relations will remain valid, and so the measures of project. Working in constant prices is simpler and involves less calculation than working in current prices; for the latter, every entry has to be adjusted for anticipated changes in the general price level.

It is quite possible, however, to work the whole project analysis in current prices. This has the advantage that all costs and benefits shown would be estimates of what the real prices will be in each year of the project. Furthermore, estimates of investment costs will be in current terms for the year in which they are expected to occur, so that the finance ministry can more easily anticipate these needs and budget the amounts necessary to finance the project on schedule.

The problem in this approach is that it involves predicting inflation rates. For items to be imported, some help is available in the World Bank report on Price Prospects for Major Primary Commodities (1982a), which is published biennially and updated in six-month intervals and includes an estimate of inflation in developed countries. For domestic inflation rates in developing countries, other sources will have to be consulted, but obtaining an estimate in which one can place even minimal confidence will be difficult, to say the least.

Even casting the project analysis in current terms may raise problems for the project analyst. Many governments have policy goals that call for greatly reduced inflation, and they cannot permit the circulation of official documents that assume rapid inflation will continue.

The mere mechanics of using current prices presents no analytical problem in project analysis, although it does complicate the computations. When we consider measures of project worth, some means of deflating future prices must be adopted for comparing future cost and benefit streams in terms that are free from the effects of general price increases.

Even when constant prices are used in the more conventional approach to project analysis, a table estimating the budgetary effects of the project in current terms that will prevail at least during the investment phase should be included either in the analysis or as a separate memorandum. It would list in current prices domestic currency needs, foreign exchange requirements, and subsidies. The finance ministry would then have better estimates to work with, and delays because of budgetary shortfalls could more easily be avoided.

PRICES FOR INTERNATIONALLY TRADED COMMODITIES

For commodities that enter significantly in international trade, whether inputs or outputs, project analysts usually obtain price information from various groups of specialists who follow price trends and make projections about relative prices in the future. In many countries where agricultural exports are important, there are groups in the agriculture ministry or the finance ministry whose help may be sought.

There are also several international organisations and trade groups to which the analyst may turn. The World Bank, for instance, publishes its projections under the title Price Prospects forMajorPrimary Commodities. The Food and Agriculture Organisation (FAO) sponsors intergovernmental groups that publish price information on rice; grains (other than rice); citrus; hard fibers; fibers (other than hard fibers); oilseeds, oils and fats; bananas;

wine and wine products; tea; meat; and cocoa. Information may be obtained from the secretary of the relevant intergovernmental group at the FAo headquarters in Rome or from the FAo representative in individual countries.

Several international commodity organisations keep detailed price information for the products of their interest. These include the International Tea Committee, the International Cocoa Organisation, the International Wool Secretariat, the International Coffee Organisation, the International Association of Seed Crushers, the International Rubber Study Group, and the International Sugar Organisation, all with headquarters in London; the International Olive Oil Council in Madrid; and the International Cotton Advisory Committee in Washington.

Some individual nations systematically collect production and price information for crops and livestock products of interest to them, and they often are willing to share this information with analysts in other countries without charge or restriction. The United States Department of Agriculture-probably the most important of these-publishes detailed studies about most major crops traded in international markets. Information may be obtained from agricultural attaches in American embassies, or directly from the department's Foreign Agriculture Service. The Commonwealth Secretariat in London publishes information about price trends for commodities of interest to its member nations.

FINANCIAL EXPORT AND IMPORT PARITY PRICES

In projects that produce a commodity significant in international trade, the price estimates are often based on projections of prices at some distant foreign point. The analyst must then calculate the appropriate price to use in the project accounts, either at the farm gate or at the project boundary.

If the farm-gate or project boundary prices for the internationally traded commodities in the project are already known, and the prices in the particular country tend to follow world market prices, the farm-gate prices may be adjusted by the same relative amount as indicated, say, by the medium trend projected in the future relative prices supplied by one or another international organisation. Also, in financial analysis, if the farm-gate price is set administratively and is not allowed to adjust freely to world prices, the relevant price to use is the administratively set price.

Table. 4.2: Elements of C.i.f. (Cost, Insurance, Freight) and F.O.B. (Free on Board).

Item	Element
C.I.F.	*Includes:*
	F.o.b. cost at point of export
	Freight charges to point of import
	Insurance charges
	Unloading from ship to pier at port

	Excludes:
	Import duties and subsidies
	Port charges at port of entry for taxes, handling, storage, agents' fees and the like
F.O.B.	*Includes:*
	All costs to get goods on board-but still in harbour of exporting country:
	Local marketing and transport costs
	Local port charges including taxes, storage, loading, fumigation, agents'
fees and	
	the like
	Export taxes and subsidies
	Project boundary price
	Farm-gate price

Simply adjusting domestic prices by the same relative amount as foreign prices often arrives at figures too rough for project analysis. The approach ignores the fact that marketing margins in commodity trade tend to be less flexible than the commodity prices themselves. There are also many instances in estimating the economic value of a traded commodity that involve deriving a shadow price based on international prices. In such instances it is necessary to calculate export or import parity prices.

These are the estimated prices at the farm gate or project boundary, which are derived by adjusting the c.i.f. (cost, insurance, and freight) or f.o.b. prices by all the relevant charges between the farm gate and the project boundary and the point where the c.i.f. or f.o.b. price is quoted in Table 4.2 previous.

One common case for which an export parity price has to be calculated is that of a commodity produced for a foreign market. Table 4.3 gives an example based on the Rahad project in the Sudan. It shows the generalized elements for calculating export parity prices so that the same methodology can be applied in other cases. As noted earlier, the Rahad project included cotton gins. Since, the gins produce lint and cottonseed for export and scarto, a by-product of very short fibers not suitable for export and sold locally, the analyst needed three prices.

For the lint and seed estimates, he began with forecasts of the 1980 c.i.f. prices in current terms at Liverpool, which were available from World Bank publications.

From these c.i.f. prices, he then deoucted insurance, ocean freight, export duties, port handling costs, and rail freight from the cotton gin at the project site to Port Sudan, thus obtaining the export parity prices at the project boundary: LSd 178.650 for lint and LSd18.097 for seed. (The symbol for Sudanese pounds is LSd.) The price for scarto, which was not exported, was based on the prevailing domestic price.

To illustrate, we may continue to calculate the export parity price at the farm gate, although in the Rahad example, where the farm-gate price was sct administratively, this calculation was not made. The computations are laid out that continues from the entry for "Equals export parity price at project boundary." Here a new issue arises. The three products that the gin produces-lint, seed, and scarto-must be converted into their seed cotton equivalents, since, it is seed cotton that the farmer sells. Similar conversions have to be made in many other instances-for example, rice milling or groundnut decortication. For the Rahad project, a weighted price of LSd83.239 forthe seed cotton was calculated using a ginning outturn of 40 percent lint, 59 percent seed, and 1 percent scarto. From this weighted price were deducted the ginning, baling, and storage charges and the costs of collection and transport from the farm gate to the gin, thus arriving at the farm-gate export parity price of LSd66.946.

Table. 4.3: Financial Export Parity Price for Cotton, Rahad Irrigation Project, Sudan (1980 forecast prices).

Step in the calculation	Relevant step in the Sudanese example	Value per ton		
		Lint	Seed	Scarto
C.i.f. at point of import	C.i.f. Liverpool (taken as estimate for all European ports)	US$639.33	US$103.39	
Deduct unloading at point of import *Deduct* freight to point of import *Deduct* insurance	Freight and insurance F.o.b. Port Sudan	39.63	24.73	
Equals f.o.b. at point of export	F.o.b. Port Sudan	US 599.70	US$ 78.66	
Convert foreign currency to domestic currency at official Exchange Rate	Converted at official Exchange Rate of LSd 1.000=US$2.872	LSd 208.809	LSd 27.389	
Deduct tariffs	Export duties	17.813	1.000	
Add subsidies *Deduct* local port charges	(None) Port handling cost Lint: LSd 5.564 per ton Seed: LSd 1.510 per ton	5.564	1.510	
Deduct local transport and marketing costs from project to point of export (if not part of the project cost)	Freight to Port Sudan at LSd 6.782 per ton	6.782	6.782	
Equals export parity price at project boundary	Export parity price at gin at project site	LSd 178.650	LSd 18.097	
Conversion allowance (if necessary)	Convert ot seed cotton (LSd 178.650 x 0.4 + LSd 18.097 x 0.59 + LSd 110.200 x 0.01)	71.460	10.677	1.102
Deduct local storage, transport, and marketing costs (if not part of project cost)	Ginning, baling, and storage (LSd 15.229 per ton)	15.229		

	Collection and internal transfer (LSd 1.064 per ton)	1.064
Equals export parity price at farm gate	Export parity price at the farm gate	LSd 66.946

LSd Sudanese pounds. US$ U.S., dollars

a. Scarto is a by-product of very short, soiled fibers not suitable for export and is sold locally at a price of LSd 110.200 per ton.

b. Seedcotton is converted into lint, seed, and scarto assuming 1 ton of seed cotton yields 400 kilograms seed, and 10 kilograms scarto.

Table. 4.4: Financial Import Parity Price of Early-crop Maize

Steps in the calculation	Relevant steps in the Nigerian example	Value per ton
F.o.b. at point of export	F.O.B. U.S., Gulf ports, No. 2 U.S., yellow corn in bulk	US$116
	Freight and insurance	31
Add freight to point of import *Add* unloading at point of import *Add* insurance *Equals* c.i.f. at point of import	C.I.F. Lagos or Apapa	US$147
Convert foreign currency to domestic currency at official Exchange Rate *Add* tariffs *Deduct* subsidies *Add* local port charges	Converted at official Exchange Ratef of N1 = US$1.62 none none Landing and port charges (including the cost of bags)	N91 22
Add local transport and marketing costs to relevant markets	Transport (based on a 350-kilometer average)	18
Equals price at market	Wholesale price	N 131
Conversion allowance if necessary	(Not necessary)	
Deduct transport and marketing costs to relevant market	Primary marketing (includes assembly, cost of bags, and intermediary margins)	-14
	Transport (based on a 350-kilometer average)	- 18
Deduct local storage, transport, and marketing costs (if not part of the project costs)	Storage loss (10 percent of harvested weight	- 9
Equals import parity price at farm gate	Imported parity price at farm gate	N 90

N = Nigerian naira

a. Forecast from World Bank, Price Prospects for Major Primary commodities.

A parallel computation leads to the import parity price. Here the issue is the price at which an import substitute can be sold domestically if it must compete with imports. Table 4.4 illustrates this issue with The example of maize production in Nigeria. The same example is presented diagrammatically. Nigeria is a net maize importer, and the project is to produce maize for domestic consumption to replacc imported maize.

We begin with the f.o.b. price at the point of export-in this case U.S., ports on the Gulf of Mexico-derived from World Bank commodity estimates. To this we add freight and insurance to obtain the c.i.f. price at either Lagos or Apapa, the two Nigerian ports concerned. Then we would add any tariffs and subsidies (in this case there are none); add local port charges for harbour dues, fumigation, handling, and the like; and add local transport to the relevant inland market. The result is the wholesale price of imported maize. It is this wholesale price of maize in the inland market that is the focal point of our calculation. The alternative to project production is not to import the maize and transport it to the project area. Rather, the alternative is to import it and market it directly on the inland market. Thus the price the farmer can expect to receive in the absence of tariffs, subsidies, or an import ban is the wholesale price less the cost of moving his maize to the market. If the project had included processing facilities, then the relevant project boundary price would have been this wholesale price less handling costs from the processing facility to the wholesale market. In the Nigerian project, no processing facilities were included, so the relevant import parity price is the farm-gate price. As we move back from the wholesale market to the farm gate, we would have to provide for any conversion allowance. In this case none is necessary, since, it is assumed that the farmer will sell shelled maize. From the wholesale price, then, we deduct local marketing costs including assembly, bags, and intermediary margins, transport from the farm to the market, and storage losses, thus obtaining the import parity price at the farm gate of N90. (The symbol for Nigerian naira is N.) This is the maximum price the farmer could expect to receive, again in the absence of tariffs, subsidies, or an import ban.

EFFECTIVENESS OF PRICE POLICIES

The next step is to examine how the objectives-constraints-policies framework can be made operational. The analytical approach views policy-makers as enacting policies (price, macro, or investment) to further objectives (efficiency, distribution, or food security) in the face of economic constraints (supply, demand, and world prices). The main services policy analysis can provide to policy-makers are to distinguish whether a policy is likely to improve the efficient operation of the economy and thus raise the level of national income, to measure the expected magnitude of the efficiency gains or losses, and to quantify, when possible, the direction and extent of the policy's likely effects on the distributional and food security objectives. Even when the non-efficiency effects are difficult to measure, economic analysis can provide a reasonable estimate of the efficiency costs associated with the promotion of non-efficiency objectives.

The ways in which agricultural price policy can lead to efficient gains are limited to offsetting market failures, assisting agricultural infant industries, and stabilising domestic prices. In developing economies, the most prevalent market failures usually are found in the factor markets, particularly for capital and occasionally for labour. These market failures are caused by insufficient development of institutions (such as financial intermediaries) and communication networks (so that information on jobs is not widespread). A second type of market failure is the existence of monopolies or monopsonies, where only one or a few (cooperating) sellers or buyers have the ability to manipulate market prices to their own advantage. Externalities (costs for which the person responsible cannot be charged or benefits that cannot be appropriated by the enterprise creating them) are a third source of market failures. Public goods are the principal source of externalities in developing countries. A public good is inadequately provided because not all of those benefiting from it can be charged for their use of it; governments thus invest in public infrastructure (roads, ports, large irrigation works), which would otherwise be inadequately supplied by private individuals.

The two other rationales for efficient intervention may also be viewed as responses to special kinds of market failures. One is to assist agricultural infant industries by correcting for dynamic market failures. The essence of the infant industry argument is that, over time, the existence of market failures (usually in the capital market or because of information bottlenecks) will cause insufficient investment and technical change and thus not permit the economy to benefit from dynamic learning effects. The presence of efficient operations in the future is not enough to justify policy that offsets the market failures; the efficiency cost to society of the inefficient use of resources in the early years must be compensated by larger efficiency gains in the later years.

The third rationale is to stabilize domestic agricultural prices (relative to unstable world prices) when insurance markets are absent. Governments perceive benefits from reducing price risks for producers, fending off consumer pressures, averting hunger if food crop prices rise, and avoiding adjustment costs for producers and consumers. Price stabilization requires public intervention in international trade and domestic marketing (transport and storage). If a public agency manages a buffer stock, the benefits from price stability may justify producer prices that are lower than average world prices and consumer prices that are higher than world prices. This margin should cover the costs of buffer stock management.

The circumstances for efficient intervention - offsetting of domestic market failures, assistance to infant industries, and stabilization of domestic prices - are potentially widespread. Analysts of efficient policy intervention look for the sources of market failure and assess the present and future benefits and costs of such policies. Even efficient intervention typically has costs as well as gains. The analyst of non-efficiency objectives begins by measuring constraints and then considers the effects of policy on objectives. If full information is available on a single commodity, the analyst can summarize the supply constraints into a supply schedule and the demand constraints into a demand schedule and can draw a standard price-quantity diagram that will portray the situation before the policy is enacted.

If a policy to raise the price of a commodity-for example, a tariff on competing imports-is put into place, the analyst examines the hypothetical effects of the restrictive trade policy on government objectives. The tariff (tax on imports) raises the domestic price to producers and consumers and influences the quantities produced, consumed, and traded internationally. Facing a higher price, producers will increase output (because they can cover higher costs of production), consumers will cut back consumption (and shift to cheaper substitutes), and the country's demand for imports of the commodity will decline on both accounts. The impact on efficiency will be negative-producers will overproduce and consumers will underconsume relative to the world price-unless the higher domestic price serves to offset a market failure. The trade policy will redistribute income, causing transfers from consumers (who will consume less at the higher price) to producers (who will grow more at the higher price) and to the government treasury (which will receive the tariff revenue on remaining imports); the effect on distribution will depend on how well off the producers and consumers are without and with the policy. The influence of the policy on food security depends on the relative stability of the additional domestic output versus the imports it replaces.

This simplified example shows how a price policy can be analyzed in order to identify its effects on government objectives. In actual price policy analysis, the process is more complicated. The first step in choosing among policies is to investigate the feasibility of the policy instrument. The imposition of a tariff on imports of a commodity can be done readily if rampant smuggling can be prevented, whereas the distribution of subsidy payments to millions of small-scale farmers might not be feasible. The next step is to measure the administrative costs of implementing the feasible instruments, for example, the costs of hiring additional customs agents. Such costs should be added to any efficiency losses of the policy or subtracted from any gains. In this way, the analyst can incorporate policy feasibility and administrative costs.

LINKAGES BETWEEN MACRO-ECONOMIC AND AGRICULTURAL PRICE POLICIES

The objectives-constraints-policies framework applies to macroe-conomic policy as well as to price policy. Common macro-economic objectives include rapid economic growth, a desirable distribution of national income, reasonably low unemployment, and moderate or low inflation. In addition to facing the sectoral constraints of supply, demand, and world prices, macro-economic planners are also confronted by a need to maintain an approximate balance in the national fiscal accounts (government revenues and expenditures) and in the foreign-exchange accounts (export earnings and foreign capital inflows versus import expenditures and foreign capital outflows). The macro-economic policies available to further these objectives in light of such constraints include fiscal and monetary policies, budgetary policies, and macro price policies influencing the foreign-exchange rate, interest rate, wage rate, and land rental rate.

The direct effects of macroeconomic policy on agricultural systems are felt through the macro price policies, especially exchange-rate policy. Fiscal and monetary policies influence agricultural systems indirectly by the interest and exchange rates. Budgetary policy-decisions on allocating both the recurrent and the capital budgets of the national government-also have indirect effects on systems, because budgetary choices influence agricultural price policy (through the availability of recurrent funds for subsidies) and public investment policy for agriculture. The three kinds of macro price policies affecting factor prices can be important in individual factor markets, although little can be said about them in general.

Some useful general lessons can be drawn from the relationships among fiscal and monetary policy, inflation, and the exchange rate and those between the exchange rate and price policies. Inflation is caused principally by macroeconomic policy-decisions to run fiscal deficits financed by expansionary monetary policy-abetted by inflation abroad that causes the prices of imports and exports to rise. If the government chooses to have a fixed-exchange-rate regime, the exchange rate will be changed only through discrete policy decisions, not because of market forces. When governments create inflation and then choose not to depreciate the nominal value of their currencies (by changing the exchange rate so that more units of domestic currency are required for each unit of foreign currency), profits are squeezed in agricultural systems that produce tradable commodities. The real exchange rate becomes overvalued when the rate of depreciation is less than the rate of inflation. Overvaluation of the real exchange rate imposes an implicit tax on producers of tradables (by keeping the domestic currency prices of their outputs artificially low), forces farmers growing tradable food crops to pay implicit food subsidies that benefit consumers, and permits artificially cheap imported inputs. A policy creating inflation with fixed nominal exchange rates squeezes agricultural profits, transfers the burden of subsidizing food from the government treasury to farmers, and makes projects based on tradable inputs appear to be more profitable than they would be if the exchange rate were set appropriately.

This state of affairs can be corrected if a government chooses to change the exchange rate. Devaluations are often difficult actions to take politically, because their short-run effects usually benefit rural inhabitants who have limited political power and harm powerful urban interest groups. Some form of foreign-exchange rationing is inevitable when the real exchange rate is overvalued, and this rationing is most often achieved by quantitive restrictions on imports that compete with domestically produced manufactures. Politically powerful urban manufacturers and their employees then shift from being supporters of devaluation to being vocal opponents of it. The prices of their products are protected from the taxing effects of overvaluation by the import quota, and the overvalued exchange rate permits them to obtain tradable inputs at artificially low prices.

PRICE POLICY GRAPHS

A set of PAMs for the country's principal representative agricultural systems provides analysts and polcy-makers with informative pictures of the existing structure of policies affecting agriculture and with a useful analytic tool for investigating the effects of future policy change. However, in most countries, there is no information base to permit construction of historical PAMs that would show changes every two or three years as trends in world or factor prices and technologies changed. Budget data might be available at best for a few systems during scattered years. But informed policy analysis requires an understanding of the recent history of policy changes as well as the detailed array of profitabilities in a given base year. This need can be met at least partially by the construction of price policy graphs.

A price policy graph is a device to permit easy visual comparisons of year-to-year movements in three price series-world prices (cif import or fob export, adjusted to a domestic wholesale market level), domestic market prices (at both the wholesale and farm levels), and domestic policy prices (guaranteed floor prices to producers and announced ceiling prices to consumers). Price policy graphs, based on annual data for fifteen to twenty years in the recent past, can be constructed for the principal agricultural commodities produced and for the main tradable inputs into agriculture. They allow a quick visual review of the pattern of price levels and price stability. If historic price policy graphs are continuously updated, they can serve as particularly useful complements to PAMs in the presentation of policy analysis.

CONCLUDING COMMENTS

Several practical lessons for practitioners emerge from this study of agricultural policy analysis. Approaches to issues and the policy agenda can be organized within the objectives-constraints-policies framework, and diagrammatic analysis can be used to identify the general direction of policy effects.

Historical perspective can be provided through a compilation of price policy graphs for the most important agricultural products and inputs. Much insight is gained from using the PAM approach to the quantitive analysis of agricultural systems. The construction of PAMs, complemented by historical price graphs, provides essential baseline information for the analysis of agricultural policy.

The standard approach to agricultural policy analysis relies on estimated elasticities of supply and demand. When policies raise or lower market prices, use of the elasticities permits the analyst to quantify changes in amounts produced and consumed; income transfers among producers, consumers, and the government treasury; and efficiency losses or gains. The PAM calculations usually are based on budget data, not elasticities. A strength of the PAM method is the disaggregation of supply in terms of technology and agroclimatic zone. Such disaggregation permits a detailed understanding of constraints on systems and provides a basis for the analysis of investment and technological change influencing the dynamic comparative advantage of agricultural systems. The principal weakness of the PAM approach is that empirical applications may not correctly specify all the marginal adjustments to alterations in output and input prices. Without sufficient information (such as elasticities of output supply and input demand), exact PAMs cannot be constructed, and approximations must be made. Unless this is done, the empirical researcher will be left with nothing more than a numberless diagram, little understanding of how the many divergences affecting agricultural systems offset one another, and no input into the policy-making process. Budget-based PAMs fill this gap in agricultural policy analysis.

MARKET STABILISATION POLICIES

Reserve Bank has proposed to the Government of India to authorise issuance of existing debt instruments, viz., Treasury Bills and dated securities up to a specified ceiling to be mutually agreed upon between the Government and the Reserve Bank by way of a Memorandum of Understanding (MoU) under the Market Stabilisation Scheme (MSS). The bills/bonds issued under MSS would have all the attributes of the existing Treasury Bills and dated securities. The bills and securities will be issued by way of auctions to be conducted by the Reserve Bank. The Reserve Bank will decide and notify the amount, tenure and timing of issuance of such treasury bills and dated securities. Whenever such securities are issued by the Reserve Bank for the purpose of market stabilisation and sterilisation, a press release at the time of issue would indicate such purpose. For the present, the total outstanding obligations of the Government by way of bills/securities thus issued under the MSS from time to time would not exceed ₹ 60,000 crore.

The bills and securities issued for the purpose of MSS would be matched by an equivalent cash balance held by the Government with the Reserve Bank. Thus, there will only be a marginal impact on revenue and fiscal deficits of the Government to the extent of interest payment on bills/securities outstanding under the MSS. Further, the cost would be shown separately in the Budget. This would add transparency to the cost of sterilisation.

MARKET STABILIZATION SCHEME (MSS)

This scheme came into existence following a MoU between the Reserve Bank of India (RBI) and the Government of India (GoI) with the primary aim of aiding the sterilization operations of the RBI.

Historically, the RBI had been sterilizing the effects of significant capital inflows on domestic liquidity by offloading parts of the stock of Government Securities held by it. It is pertinent to recall, in this context, that the assets side of the RBI's Balance Sheet (July 1 to June 30) includes Foreign Exchange Reserves and Government Securities while liabilities are primarily in the form of High Powered Money (consisting of Currency with the public and Reserves held in the RBI by the Banking System). Thus, any rise in Foreign Exchange Reserves resulting from the intervention of the RBI in the FOREIGN EXCHANGE MARKETS (with the intention, say, to maintain the exchange rate on the face of huge capital inflows) entails a corresponding rise in High Powered Money. The Money Supply in the economy is linked to High Powered Money via the money multiplier. Therefore, on the face of large capital inflows, to keep the liabilities side constant so as to not raise the Supply of Money, corresponding reduction in the stock of Government Securities by the RBI is necessary.

The MSS was devised since, continuous resort to sterilization by the RBI depleted its limited stock of Government Securities and impaired the scope for similar interventions in the future.

Under this scheme, the GoI borrows from the RBI (such borrowing being additional to its normal borrowing requirements) and issues Treasury-Bills/ Dated Securities that are utilized for absorbing excess liquidity from the market. Therefore, the MSS constitutes an arrangement aiding in liquidity absorption, in keeping with the overall monetary policy stance of the RBI, alongside tools like the Liquidity Adjustment Facility (LAF) and Open Market Operations (OMO).

The securities issued under MSS, termed as Market Stabilization Scheme (MSS) Securities/Bonds, are issued by way of auctions conducted by the RBI and are done according to a specified ceiling mutually agreed upon by the GoI and the RBI. They possess all the attributes of existing Treasury-Bills/Dated Securities and are included as a part of the country's 'internal Central Government debt'.

The amount raised under the MSS does not get credited to the Government Account but is maintained in a separate cash account with the RBI and are used only for the purpose of redemption/buy back of Treasury-Bills/Dated Securities issued under the scheme.

However, following the global financial crisis of 2008, that necessitated fiscal stimulus measures, an amendment to the original MoU between the RBI and the GoI in February 2009 allowed the Government to convert a portion of the MSS funds into normal government borrowing for FINANCINGf**** its stimulus expenditure requirements.

Treasury-Bills/Securities issued under MSS are matched by equivalent cash balances that are held by the Government with the RBI. Such payments are not made from the MSS account just as receipts due to premium or accrued interest on these Securities are not credited to it. As and when MSS securities are issued by the RBI as well as the annual ceiling, when decided, is notified through a press release. For the fiscal year 2010-11 the annual ceiling for such securities outstanding stand at ₹ 50,000 crore, with a review due when the outstanding reaches the threshold of ₹ 35,000 crore.

STABILISATION POLICY

A stabilisation policy is a package or set of measures introduced to stabilize a financial system or economy. The term can refer to policies in two distinct sets of circumstances: business cycle stabilisation and crisis stabilisation. In either case, it is a form of discretionary policy.

BUSINESS CYCLE STABILISATION

Stabilisation can refer to correcting the normal behaviour of the business cycle. In this case the term generally refers to demand management by monetary and fiscal policy to reduce normal fluctuations and output, sometimes referred to as "keeping the economy on an even keel."

The policy changes in these circumstances are usually countercyclical, compensating for the predicted changes in employment and output, to increase short-run and medium run welfare.

CRISIS STABILISATION

The term can also refer to measures taken to resolve a specific economic crisis, for instance an exchange-rate crisis or stock market crash, in order to prevent the economy developing recession or inflation.

The package is usually initiated either by a government or central bank, or by either or both of these institutions acting in concert with international institutions such as theInternational Monetary Fund (IMF) or the World Bank. Depending on the goals to be achieved, it involves some combination of restrictive fiscal measures (to reduce government borrowing) and monetary tightening (to support the currency).

- Recent examples of such packages include Argentina's re-scheduling of its international obligations (where central banks and leading international banks re-scheduled Argentina's debt so as to allow it to avoid total default), and IMF interventions in South East Asia (at the end of the 1990s) when several Asian economies encountered financial turbulence.

This type of stabilisation can be painful, in the short term, for the economy concerned because of lower output and higher unemployment. Unlike a business-cycle stabilisation policy, these changes will often be pro-cyclical, reinforcing existing trends. While this is clearly undesirable, the policies are designed to be a platform for successful long-run growth and reform.

It has been argued that, rather than imposing such polices after a crisis, the international financial system architecture needs to be reformed to avoid some of the risks (*e.g.*, hot money flows and/or hedge fund activity) that some people hold to destabilize economies and financial markets, and lead to the need for stabilisation policies and, *e.g.*, IMF interventions. Proposed measures include for example a global Tobin tax on currency trades across borders.

HOW MARKET INTELLIGENCE CAN HELP PRICE STABILISATION FUND

The recent plunge in potato prices to 3-4 a kg in West Bengal has once again raised serious concern over marketing related policies of government. It is not the absence of support which makes farmers so vulnerable; but it is the approach of the policy to address the issue which matters. After farmers committed suicide, the West Bengal Government announced its plan to procure the tuber at a price higher than what market was offering.

The initiative of the State government to procure potato at 5/kg has arrested further fall in price.

However, the short-term approach is not the solution to farmers getting remunerative price. Incidences of farmers' committing suicide in West Bengal have taken place at a time when the Centre is finalising the guidclincs for a Price Stabilisation Fund (PSF) which aims to provide support to State government in procuring perishables such as potato and onion when farmers are not getting remunerative price .

Why Price Falls

Seasonality is an inherent factor with the production of agricultural commodities, which often results in price turning volatile. But the interesting factor is that many of the agricultural commodities do not reflect the true fundamental picture. The price discovered for agricultural products are reflection of demand and supply equation confined to a particular geographical area only.

While the average weekly price of potato during second week of March in West Bengal was 415.48 a quintal, at the same time prices were ruling higher in Kerala (2,296), Karnataka (1,500), Meghalaya (1,267), Andhra Pradesh (1,239), Tripura (1,113), Maharashtra (994), Telangana (824).

The spatial variation in price across various markets in the country indicates that there are some factors prohibiting price rise by considering the demand and supply situation in other States as well. Hindrances in inter-State movement of agricultural goods have lot to do with this price variation, exaggerated by the move of State government to ban the inter-State trade of potato two years ago. It is high time that the States came forward in amending their agricultural marketing policy to support the Centre in establishing a barrier-free market.

INFORMATION ASYMMETRY

PSF is an important step to safeguard the interest of producers and consumers against price volatility that has been observed in the case of horticultural commodities in the recent past, mainly onion and potato. Availability of stocks has a major impact on prices stabilising as has been envisaged through this scheme on price stabilisation.

Considering major factors which may vary from season to season, the area under cultivation and climatic variables play an important role in defining total production in a particular region. Improvement in available infrastructure and various government and non-government machinery is a prerequisite for real time collection, dissemination and assessment of data pertaining to these variables. Currently, agricultural data's reliability and on time availability have been challenged. With present guidelines of the PSF, farmers will be offered Government support only after they face hardship. But again, the approach can be even more efficient by taking cognisance of farmer's requirement well before sowing begins. The obsolete method of data collection, projection and analysis needs to be replaced with advanced technology, which is faster and reliable. PSF should include a component for strengthening of data collection and dissemination method.

This may result into not only strengthening of information system, but will also be a facilitating factor in decision making.

MARKET INTELLIGENCE

Advance indication of prices help farmers in making better decision for their land utilisation. Currently, information is disseminated to farmers through Price Ticker Board (PTB). Farmers refer to these PTBs only when they go to APMCs to sell their produce. What farmers need is advance price information to facilitate sowing decision.

The scheme needs to be linked with other schemes of the Department such as Agri Tech Infrastructure Fund. This inter-linkage will ensure availability of sufficient infrastructure for efficient supply chain and better price discovery.

STABILISATION METHODS

The three main methods of publicly assisted stabilisation measures are through:

- Short term market management using intervention purchasing and border measures.
- Medium term market management using supply control schemes (*e.g.* quotas and set-aside).
- Revenue or income insurance schemes.

From the EU perspective, there is a strong presumption in favour of the first two approaches. The Union has the instruments and a great deal of experience in using them. However, if it is decided to continue with these tried and tested instruments, it is vitally important that there are built-in safeguards to ensure that stabilisation will not slip again into systematic support and protection. It is instructive that in the 1996 US Farm Bill the instruments of set aside was abolished altogether to remove any possibility that it could slip back into use. One response is to amend regulations so that the price stabilisation objectives are clear, and also to redefine intervention prices (there is a different concept used for different commodities) clearly in relation to world prices. For example, the single grains intervention price could be a fixed proportion of the moving average of the international price for grains. The intention should be that market intervention is an exceptional event, taking place only rarely to limit the worst falls in price.

A variant of this approach would be to devolve responsibility for price stabilisation to an agency which only has this duty, and is not empowered to do anything other than manage public intervention stocks to reduce price fluctuations. Such an agency could, as far as possible, work through the private market, empowering or stimulating private storers to carry out the necessary intervention, as it is already used in various market organisations. To the extent that this could be done, it would help minimise the public costs of maintaining storage and handling capacity which, by definition, will only be used infrequently.

Border measures can play a part in domestic stabilisation. For many years, the EU used the instruments of variable import levies and variable export refunds as primary tools for market stabilisation. Undoubtedly, these were major contributors to domestic market stability. However, this was at the expense of international market instability, and it was of course agreed in the Uruguay Round that variable levies should be converted to tariffs and the use of export subsidies reduced. In recognition of the right of countries to defend their domestic markets against sudden surges of imports or collapse in import prices, the URAA allows the use of safeguards, *i.e.,* higher tariffs, in well defined ways. This is, thereforem an internationally agreed tool of market price stabilisation.

There is concern in some quarters that longer term stabilisation instruments may be necessary for some markets. If world grain prices again become seriously depressed to the extent that a significant section of European grain production is threatened, then to avoid the building up of intervention stocks and the system switching, once again, from short term stabilisation to longer term market support, it might be justifiable to re-introduce land set-aside or other supply management tools. Clearly, set-aside has no place in short term market management; it is simply too slow to take land out of production next year to solve a price slump this year. Likewise, once the land is fallowed, there is nothing which can be done, within a period of a year, to increase supplies to deal with a shortfall.

There is also a danger that the slow-acting set-aside could exacerbate an unfavourable market development. There are no clearer examples of this than the 1995 decisions by the US to idle 7.5 per cent of their corn area and the EU to have a 12 per cent COP set-aside. These coincided with poor weather conditions in several markets and contributed to the 1996 shortage and price surge. The main burden of these 'wrong' decisions is of course borne by importing countries which include some of the poorest areas of the world.

These longer term considerations raise the issue of who is going to take responsibility for the third objective of stabilisation, namely, concerns with international food security and specifically, global stock management. Some would answer, "no one", "the market is the best device for balancing supply and demand". As both US and EU public stocks have all but vanished, and neither shows a great desire to take on this global role, this perhaps looks the most likely choice. However, it is not a satisfactory answer. It is plain that neither of the two largest players in international grain markets (the US and the EU) have any intention of leaving their domestic markets entirely free of intervention. At present, the US is taking, and the EU is talking about taking, a minimalist approach, designed for relatively modest domestic stabilisation. But, such is the importance of these two blocs in world grain trade, that this is not likely to be the end of the story. Each will continue to see a role for some stocks, at the very least for famine relief. Beyond this, the situation is quite unclear.

A completely different approach to stabilisation is through income or revenue insurance schemes. This recognises that price stability *per se* is not the problem. The most important social cost of instability is that, in a bad slump, many producers are driven out of business. This is bad for them, it is costly because it creates economic and social disruption for other businesses which traded with the failed firms, and the threat of such failures encourages a more conservative approach to investment than is socially desirable. In short, the problem occurs when the revenues of farms fall below their survival threshold.

Superficially, it could be argued that the MacSharry reforms, *de facto,* introduced a major element of revenue or income assurance through the arable and livestock compensatory payments. These payments certainly provide a strong degree of income stability. In the case of the Spanish grain sector in 1994 and 1995, many farmers were saved by these payments given the severe drought in those years. However the origins, recipients and basis of the compensation payments contain none of the desirable elements of risk insurance. There are no premiums paid for the insurance, no definition of the risk insured, and no mechanism to ensure that only victims are compensated. With the arable compensation payments, all grain and oilseed farmers get the payments irrespective of whether their revenues have fallen or risen, and there is no mechanism to ensure that farmers of other crops affected by drought receive assistance. The arable payments were not set up to deal with risk management so it is only accidental if they can help.

There are potentially strong advantages in turning from price stabilisation to revenue stabilisation. It allows for the fact that most farms are multi-product businesses and it is unusual that all markets fail simultaneously, and it also allows for the inverse relation between prices and quantities (when prices are low it is usually because quantities are high and *vice versa*). It is focused on the real problem caused by instability, and it pays out only to those in need of assistance.

A move to revenue insurance would also have the advantage of enabling a complete withdrawal of the 'state' from all market management. Despite these attractions of the insurance approach, it is not used as a central part of agricultural policy in any country. Even where it is used, it is confined to the producers of the major field crops and is not a general facility available to all farmers. The principal problems are those of moral hazard. This has several dimensions. First, there is an asymmetry of information, the insured usually know more about their precise conditions, their actions and the risks, than the insurer. Second, there is a tendency for adverse selection; insurance schemes will naturally attract more of the high risk farmers. Third, if farmers are insured, there is a danger that they will be more careless about risks than otherwise, thereby raising the overall costs of providing the insurance.

For these reasons, the private insurance sector will generally not get involved in this business. This means that such schemes must be state supported. That in itself is not an objection; after all, the present support arrangements done in the name of stabilisation are very heavily state supported. However, governments themselves seem reluctant to move into the insurance business, and especially on behalf of just one part of one economic sector (grain farmers). Other farmers and some other small businessmen face just as risky circumstances without such state revenue insurance.

It would be a major departure for the EU to move in this direction, but it is certainly worth close examination to see whether it could play a role under a market stabilisation programme. At the same time, to assure farmers that a significant reduction in market price support does not mean that the authorities are abandoning farmers completely to the market, it would be sensible to follow the suggestions of the 1995 Spanish Presidency suggestions to review the desirability of harmonising member state national insurance and disaster relief schemes and to consider if there is a role for EU action in either of these. These matters are presently handled mostly by Member States, although from time to time the EU is called to assist with natural disasters, *e.g.,* floods or drought and, most recently.

POLICY ANALYSIS MATRIX IN AGRICULTURE

This section explains the construction of the policy analysis matrix and the derivation of measures of efficiency and policy transfer used in agricultural policy analysis. The study of agricultural policy spans three levels-microeconomic behaviour of producers, marketing and trade, and macroeconomic linkages. Practitioners of agricultural economics typically give different emphasis to these three topics; micro production issues receive the greatest attention, marketing and trade get less, and macroeconomic links receive little or no coverage. This book argues that excessive specialisation precludes successful policy analysis; applied agricultural economists need to understand all of the components of and links among farming systems, domestic and international markets, and macroeconomic policy. Policy analysts have to appreciate feedbacks and tradeoffs within the big picture.

The PAM approach is a system of double-entry book - keeping. Analysts using PAM have to provide complete and consistent coverage to all policy influences on returns and costs of agricultural production. With this method, applied economists need to be equally capable of analysing, for example, fertilizer response functions, quantitative restrictions on trade, and real effective exchange rates. The main empirical task is to construct accounting matrices of revenues, costs, and profits. A PAM is constructed for the study of each selected agricultural system-using data on farming, farm-to-processor marketing, processing, and processor-to-wholesaler marketing. The impact of commodity and macroeconomic policies can then be gauged by comparison with the absence of policy.

PRACTICAL ISSUES ADDRESSED

Three principal issues - the impact of policy on competitiveness and farm-level profits, the influence of investment policy on economic efficiency and comparative advantage, and the effects of agricultural research policy on changing technologies - can be investigated with the PAM approach. The results can be used to identify what kinds of farmers - categorized by the commodities they grow, the technologies they use, and the agroclimatic zones in which their farms are located are competitive under current policies affecting crop and input prices and how their profits change as the policies are altered. This issue of farm policy-how agricultural prices affect farming profits - is of primary importance to ministries of agriculture. In the PAM approach, farm budget data (sales revenues and input costs) are collected for the principa, agricultural systems. The determination of profit actually received by farmers is a straightforward and important initial result of the analysis. It shows which farmers are currently competitive and how their profits might change if price policies were changed.

A second issue concerns the economic efficiency (or comparative advantage) of agricultural systems and how additional public investment might change the current pattern of efficiency. In what commodity production systems, defined by technology and agroclimatic zone, does the country currently exhibit strong or weak comparative advantage, and how might new investments, using government revenues or foreign aid funds, improve this picture? Investment policy is of primary interest to economic planners who allocate capital budgets, including foreign aid, in attempts to increase efficiency and speed the growth of national income.

With the PAM method, the analyst reassesses the revenues, costs, and profits indicated in farm-level and marketing budgets. Efficiency valuations of outputs and inputs are meant to lead to the highest possible levels of national income. The difference between revenues and costs for a system-both valued in social prices-is social profits, a measure of economic efficiency. New investments that reduce social costs also increase social profits and improve efficiency. An understanding of the array of social profitabilities of agricultural systems greatly reduces the number of detailed benefit-cost analyses needed to evaluate investment alternatives.

A third and closely related set of issues is how best to allocate funds for agricultural research. How can economic analysis be used to help determine the most fruitful directions for primary and applied research to raise crop yields and reduce social costs, thereby increasing social profits? This question is faced by decision-makers in the international agricultural research centers, in several international organisations, and in the agricultural research establishments of certain countries. It is a question also asked by central planners who make allocations to agricultural research budgets.

The approach used in PAM analysis begins with the calculation of existing levels of private (actual market) and social (efficiency) revenues, costs, and profits. This calculation reveals the extent to which actual profits are generated by policy transfers rather than by underlying economic efficiency. Next, agricultural scientists need to project changes in yields and inputs resulting from alternative research programmes. The effectiveness of such changes can then be gauged by an examination of how they alter private and social profits of current technologies.

THE POLICY ANALYSIS MATRIX

The policy analysis matrix is a product of two accounting identities, one defining profitability as the difference between revenues and costs and the other measuring the effects of divergences (distorting policies and market failures) as the difference between observed parameters and parameters that would exist if the divergences were removed. By filling in the elements of the PAM for an agricultural system, an analyst can measure both the extent of transfers occasioned by the set of policies acting on the system and the inherent economic efficiency of the system.

Profits are defined as the difference between total (or per unit) sales revenues and costs of production. This definition generates the first identity of the accounting matrix. In the PAM, profitability is measured horizontally, across the columns of the matrix (Table 2.1).

Table.2.1: Policy Analysis Matrix

	Revenues	Costs		Profit
		Tradable Inputs	Domestic Factors	
Private Prices	A	B	C	D
Social Prices	E	F	G	H
Divergences	I	J	K	L

Notes: *Private profits*, D, equal A minus B minus C. *Social profits*, H, equal E minus F minus G. 'Output transfers, 1, equal A minus E. 1nput transfers, J, equal B minus F. *Factor transfers*, K, equal C minus G. Net transfers, L, equal D minus H; they also equal I minus J minus K.

Ratio Indicators for Comparison of Unlike Outputs:

Private cost ratio (PCR): C/(A - B). Domestic resource cost ratio (DRC): G/(E - F) Nominal protection coefficient (NPC) on tradable outputs (NPCO): A/E on tradable inputs (NPCI): B/F Effective protection coefficient (EPC): (A - B)/(E - F) Profitability coefficient (PC): (A - B - C)/(E - F - G) or D/H Subsidy ratio to producers (SRP): L/E or (D - H)/E in social evaluations), and non-tradable inputs (which themselves have to be further disaggregated so that ultimately all component costs are classified as tradable inputs, domestic factors, or transfers).

Profits, shown in the right-hand column, are found by the subtraction of costs, given in the two middle columns, from revenues, indicated in the left-hand column. Each of the column entries is thus a component of the profits identity-revenues less costs equals profits.

Each PAM contains two cost columns, one for tradable inputs and the other for domestic factors. Intermediate inputs-including fertilizer, pesticides, purchased seeds, compound feeds, electricity, transportation, and fuel-are divided into their tradable-input and domestic factor components. This process of disaggregation of intermediate goods or services separates intermediate costs into four categories-tradable inputs, domestic factors, transfers (taxes or subsidies that are set aside.

An example illustrates the process of disaggregating intermediate goods or services. Fertilizer is for most countries a tradable intermediate input. If a particular country is a net importer of fertilizer, the social valuation of a specific kind of fertilizer for its agricultural system is given by the cif (costs, insurance, freight) import price for that fertilizer plus the social costs of moving the input to the representative location in the system. Finding the import price is usually straightforward. Finding the social valuation of the domestic marketing costs is another story, however. It is necessary to study the transportation industry-road or rail-and disaggregate the costs into labour, capital, fuel, and so forth. Each type of cost then needs to be further broken down through use of an appropriate world price and an estimate of local transportation costs.

PRIVATE PROFITABILITY

The term 'private' refers to observed revenues and costs reflecting actual market prices received or paid by farmers, merchants, or processors in the agricultural system. The private, or actual, market prices thus incorporate the underlying economic costs and valuations plus the effects of all policies and market failures. The private profits, D, are the difference between revenues (A) and costs (B + C); and all four entries in the top row are measured in observed prices. The calculation begins with the construction of separate budgets for farming, marketing, and processing. The components of these budgets are usually entered in PAM as local currency per physical unit, although the analysis can also be carried out using a foreign currency per unit.

The private profitability calculations show the competitiveness of the agricultural system, given current technologies, output values, input costs, and policy transfers. The cost of capital, defined as the pretax return that owners of capital require to maintain their investment in the system, is included in domestic costs (C); hence, profits (D) are excess profits-above-normal returns to operators of the activity. If private profits are negative (D < 0), operators are earning a subnormal rate of return and thus can be expected to exit from this activity unless something changes to increase profits to at least a normal level (D = 0). Alternatively, positive private profits (D > 0) are an indication of supernormal returns and should lead to future expansion of the system, unless the farming area can not be expanded or substitute crops are more privately profitable.

SOCIAL PROFITABILITY

The second row of the accounting matrix utilizes social prices. These valuations measure comparative advantage or efficiency in the agricultural commodity system. Efficient outcomes are achieved when an economy's resources are used in activities that create the highest levels of output and income. Social profits, H, are an efficiency measure because outputs, E, and inputs, F + G, are valued in prices that reflect scarcity values or social opportunity costs. Social profits, like the private analogue, are the difference between revenues and costs, all measured in social prices-H = (E - F - G).

For outputs (E) and inputs (F) that are traded internationally, the appropriate social valuations are given by world prices-cif import prices for goods or services that are imported or fob export prices for exportables. World prices represent the government's choice to permit consumers and producers to import, export, or produce goods or services domestically; the social value of additional domestic output is thus the foreign exchange saved by reducing imports or earned by expanding exports (for each unit of production, the cif import or fob export price). Because of global output fluctuations or distorting policies abroad, the appropriate world prices might not be those that prevail during the base year chosen for the study. Instead, expected long-run values serve as social valuations for tradable outputs and inputs.

The services provided by domestic factors of production-labour, capital, and land - do not have world prices because the markets for these services are considered to be domestic. The social valuation of each factor service is found by estimation of the net income forgone because the factor is not employed in its best alternative use. This approach requires the commodity systems under analysis to be excluded from social factor price determination. For example, if land is planted to wheat, it cannot grow barley during the identical crop season; the social opportunity cost of the land for the wheat system is thus the net income lost because the land cannot produce barley. Similarly, the labour and capital used to produce wheat cannot simultaneously provide services elsewhere in agriculture or in other sectors of the economy. Their social opportunity costs are measured by the net income given up because alternative activities are deprived of the labour and capital services applied to wheat production.

The practice of social valuation of domestic factors begins with a distinction between mobile and fixed factors of production. Mobile factors, usually capital and labour, are factors that can move from agriculture to other sectors of the economy, such as industry, services, and energy. For mobile factors, prices are determined by aggregate supply and demand forces. Because alternative uses for these factors are available throughout the economy, the social values of capital and labour are determined at a national level, not solely within the agricultural sector. Actual wage rates for labour and rates of return to capital investment are therefore affected by a host of policies, some of which may distort factor prices directly. An enforced and binding minimum-wage law, for example, raises the market wage above what it would have been in the absence of policy and causes observed wages to be higher than the social opportunity cost of labour. But indirect effects can also be important. Distortions of output prices cause different activities to expand or contract, altering in turn the demand and prices of mobile domestic factors.

Fixed, or immobile, factors of production are the factors whose private or social opportunity costs are determined within a particular sector of the economy. The value of agricultural land, for example, is usually determined only by the land's worth in growing alternative crops. Because land is immobile, its value is not directly affected by events in the industrial and service sectors of the economy. But the social opportunity cost of farmland is sometimes difficult to estimate. Within any agroclimatic zone, complete specialisation in the most profitable crop is rarely observed. Instead, farmers prefer rotations or intercropping systems that reduce risks of income losses from price variability, yield losses, and pest and disease infestation. Therefore, the social opportunity cost of the land is not accurately approximated by the net profitabilities of a single best alternative crop; instead, it is measured by some weighted average of the social profits accruing from the set of crops planted. Because the correct weights and social profits associated with each crop in the set are generally not known, it is convenient in assessing farming activities to reinterpret crop profits as rents to land and other fixed factors (for example, management and the ability to bear risk) per hectare of land used. This reinterpretation includes private (and social) returns to land as parts of D (and H). Profitability per hectare is then interpreted as the ability of a farming activity to cover its long-run variable costs, in either private or social prices or as a return to fixed factors such as land, management skill, and water resources.

EFFECTS OF DIVERGENCES

The second identity of the accounting matrix concerns the differences between private and social valuations of revenues, costs, and profits. For each entry in the matrix-measured vertically-any divergence between the observed private (actual market) price and the estimated social (efficiency) price must be explained by the effects of policy or by the existence of market failures.

Table.2.2: Expanded Policy Analysis Matrix

	Revenues	Costs		Profits
		Tradable Inputs	Domestic Factors	
Private Prices	A	B	C	D
Social Prices	E	F	G	G
Diverges and efficient policy	I	J	K	L
Effects of market failures	M	N	O	P
Effects of distorting policy	Q	R	S	T
Effects of efficient policy	U	V	W	X

Notes: *Private profits,* D, equal A minus B minus C. *Social profits,* H, equal E minus F minus G. [3]Output transfers, 1, equal A minus E; they also equal M plus Q plus U. lnput transfers, J, equal B minus F; they also equal N plus R plus V. Factor transfers, K, equal C minus G; they also equal O plus S plus W. Net transfers, L, equal D minus H; they also equal I minus J minus K; and they equal P plus T plus X.

This critical relationship follows directly from the definition of social prices. Social prices correct for the effects of distorting policies-policies that lead to an inefficient use of resources. These policies often are introduced because decision-makers are willing to accept some inefficiencies (and thus lower total income) in order to further non-efficiency objectives, such as the redistribution of income or the improvement of domestic food security. In this circumstance, assessing the tradeoffs between efficiency and non-efficiency objectives becomes a central part of policy analysis.

Some expended Policy Analysis are shown in Table. 2.2 Previous.

But not all policies distort the allocation of resources. Some policies are enacted expressly to improve efficiency by whenever monopolies or monopsonies (seller or buyer control over market prices), externalities (costs for which the imposer cannot be charged or benefits for which the provider cannot receive compensation), or factor market imperfections (inadequate development of institutions to provide competitive services and full information) prevent a market from creating an efficient allocation of products or factors. Hence, one needs to distinguish distorting policies, which cause losses of potential income, from efficient policies, which offset the effects of market failures and thus create greater income. Because efficient policies correct divergences, they reduce the differences between private and social valuations.

Interpretation of the effects of divergences can be clarified by the expansion of the PAM to include six rows. In this expanded PAM, each entry measuring the effects of divergences (I, J, K, and L) is disaggregated into three categories-market failures (fourth row), distorting policies (fifth row), and efficient policies (sixth row). The introduction of efficient policies to offset market failures would change the entries in the first and third rows. To bring about perfect efficiency, a government would introduce efficient policies to offset the effects of market failures and avoid distorting policies, thereby ensuring equality of private and social prices.

In the absence of market failure in the product markets, all divergences between private and social prices of tradable output and inputs are caused by distorting policy. Because the principles are identical for all tradable products, the matrix entries for revenues (tradable outputs)and tradable inputs can be considered together. Output transfers, I = (A - E), and input transfers, J = (B - F), arise from two kinds of policies that cause divergences between observed and world product prices: commodity-specific policies and exchange-rate policy.

Policies that apply to specific commodities include a wide range of taxes or subsidies and trade policy. For example, producer revenues per unit can be raised by producer subsidies (sometimes called deficiency payments in agriculture), tariffs or import quotas on outputs (which raise domestic prices), or domestic price supports enforced by government stockpiling (which require a complementary trade restriction for tradable products). Commodity-specific policies on inputs also affect private profitability. For example, per unit producer costs can be lowered by direct input subsidies or by subsidies on imported inputs.

Typically, PAM accounting is done in domestic currency, but world prices are quoted in foreign currency. Hence, a foreign exchange rate is needed to convert world prices into domestic equivalents. The social exchange rate may differ from observed exchange rates. Undervalued exchange rates reflect an excess supply of foreign exchange that is accumulating as excessive reserves and reducing potential income. Overvalued exchange rates correspond to conditions of excess demand; this demand results in extra foreign borrowing, excessive drawing down of exchange reserves, or rationing of foreign exchange among domestic users.

An overvalued exchange rate is an implicit tax on producers of tradable products because too little domestic currency is earned by exports or paid out for imports. In the absence of commodity policy, the world price of a tradable good determines its domestic price. When the exchange rate is overvalued, the domestic price is lower than its efficiency level and domestic producers are effectively taxed. Undervalued exchange rates exert the opposite effects. Correction for this distortion in PAM is done by conversion of world prices (E and F in the matrix) at the social exchange rate rather than at the official rate. Because exchange rates affect both product prices and factor prices, exchange-rate adjustments are limited to special circumstances-the appearance of multiple exchange-rate regimes or the government's failure to adjust the exchange rate enough to offset the effects of domestic inflation.

The social costs of domestic factors (G) reflect underlying supply and demand conditions in domestic factor markets. Factor prices are thus influenced by the prevailing set of macroeconomic and commodity price policies. In addition, the government can affect factor costs with tax or subsidy policies for one or more of the factors (capital, labour, or land) that create a divergence between private costs (C) and social costs (G). Finally, market imperfections, arising from imperfect information or underdeveloped institutions-which are often characteristic of developing country economies-further influence factor prices. If factor market imperfections exist along with distorting factor policy, both O and S and possibly W are positive components of K. The net transfer, L, thus combines the effects of distorting policy (I, J, and the S part of K) with those of factor market failures (the O part of K) and efficient policies to offset them (the W part of K).

The net transfer caused by policy and market failures (L in the matrix) is the sum of the separate effects from the product and factor markets, L = (I - J - K). (Positive entries in the two cost categories, J and K, represent negative transfers because they reduce private profits, whereas negative entries in J and K represent positive transfers; hence, J and K are subtracted from I, a positive transfer, in the calculation of the net transfer, L.) The net transfer from distorting policy is the sum of all factor, commodity, and exchange-rate policies (apart from efficient policies that offset market failures).

The net transfer can also be found by a comparison of private and social profits. These measures of the net transfer must by definition be identical in the double-entry accounting matrix, L = (I - J - K) - (D - H). Disaggregation of the total net transfer shows whether each distorting policy provides positive or negative transfers to the system. The PAM thus permits comparison of the effects of market failures and distorting policies for the entire set of commodity and macroprice (factor and exchange-rate) policies. This comparison can be made for the complete agricultural system and for each of its outputs and inputs.

COMPARISONS AMONG AGRICULTURAL SYSTEMS PRODUCING DIFFERENT OUTPUTS

The entries in PAM allow comparisons among agricultural systems that produce identical outputs, either within a single country or across two or more countries. In the accounting matrix, all measures are given as monetary units per physical unit of some commodity. If interest focuses solely on a comparison of one wheat system with another, for example, the matrix entries provide all information necessary for the analysis. Comparisons can be drawn readily by construction of PAM entries for two or more different systems that produce the same quality of wheat. (If necessary, premiums or discounts can be used to correct for quality differences.) Further comparisons can be made between the wheat systems in one country and those in other wheat-producing countries; social exchange rates, incorporating corrections for differential inflation not otherwise offset by exchange-rate changes, are used to convert the other countries' currencies into domestic currency.

Comparisons between wheat and barley-or apples and oranges are another story, however. To permit comparisons among systems producing different outputs, some common numeraire must be generated. One technique involves the expression of all values relative to a constraining domestic factor resource, such as land. A more common method uses ratios. Both the numerator and the denominator of each ratio are PAM entries defined in domestic currency units per physical unit of the commodity. Therefore, the ratio is a pure number free of any commodity or monetary designation.

PRIVATE PROFITABILITY

For comparisons of systems producing identical outputs, private profits, D = (A - B - C), indicate competitiveness under existing policies. Construction of a ratio is required to permit comparisons among systems producing different commodities. Direct inspection of the data for private profits is not sufficient. Profitability results are residuals and might have come from systems using very different levels of inputs to produce outputs with widely varying prices. This difficulty might.not be apparent in a wheat versus corn example, but it would arise in a comparison of a wheat system with one producing a high-value crop, such as strawberries. This ambiguity is inherent in comparisons of private profits of systems producing different commodities with differing capital intensities.

The problem is circumvented, by construction of a private cost ratio (PCR)- the ratio of domestic factor costs (C) to value added in private prices (A - B); that is, PCR = C/(A - B). Value added is the difference between the value of output and the costs of tradable inputs; it shows how much the system can afford to pay domestic factors (including a normal return to capital) and still remain competitive-that is, break even after earning normal profits, where (A - B - C) = D = 0. The entrepreneurs in the system prefer to earn excess profits (D > 0), and they can achieve this result if their private factor costs (C) are less than their value added in private prices (A - B). Thus they try to minimize the private cost ratio by holding down factor and tradable input costs in order to maximize excess profits.

SOCIAL PROFITABILITY

Social profits measure efficiency or comparative advantage. For a comparison of identical outputs, results can be taken directly from the second row of the PAM matrix-social profits equal social revenues less social costs, H = (E - F - G). When social profits are negative, a system cannot survive without assistance from the government. Such systems waste scarce resources by producing at social costs that exceed the costs of importing. The choice is clear for efficiency-minded economic planners: enact new policies or remove existing ones to provide private incentives for systems that generate social profits, subject to non-efficiency objectives.

When systems producing different outputs are compared for relative efficiency, the domestic resource cost ratio (DRC), defined as G/(E - F), serves as a proxy measure for social profits. No new information beyond social revenues and costs is required to calculate a DRC. The DRC plays the same substitute role for social profits as does the PCR for private profits; in both instances, the ratio equals 1 if its analogous profitability measure equals 0. Minimizing the DRC is thus equivalent to maximizing social profits. In cross-commodity comparisons, DRC ratios replace social profit measures as indicators of relative degrees of efficiency.

POLICY TRANSFERS

Transfers are shown in the third row of the PAM. If market failures are unimportant, these transfers measure mainly the effects of distorting policy. Efficient systems earn excess profits without any help from the government, and subsidizing policy (L > 0) increases the final level of private profits. Because subsidizing policy permits inefficient systems to survive, the consequent waste of resources needs to be justified in terms of non-efficiency objectives.

Comparisons of the extent of policy transfers between two or more systems with different outputs also require the formation of ratios (for reasons analogous to those offered in the discussions of private and social profits). The nominal protection coefficient (NPC) is a ratio that contrasts the observed (private) commodity price with a comparable world (social) price. This ratio indicates the impact of policy (and of any market failures not corrected by efficient policy) that causes a divergence between the two prices. The NPC on tradable outputs (NPCO), defined as A/E, indicates the degree of output transfer; for example, an

NPC of 1.10 shows that policies are increasing the market price to a level 10 percent higher than the world price. Similarly, the NPC on tradable inputs (NPCI), defined as B/F, shows the degree of tradable input transfer. An NPC on inputs of 0.80 shows that policies are reducing input costs; the average market prices for these inputs are only 80 percent of world prices.

The effective protection coefficient (EPC), another indicator of incentives, is the ratio of value added in private prices (A - B) to value added in world prices (E - F), or EPC = (A - B)/(E - F). This coefficient measures the degree of policy transfer from product market-output and tradable-input-policies. But, like the NPC, the EPC ignores the transfer effects of factor market policies. Hence, it is not a complete indicator of incentives.

An extension of the EPC to include factor transfers is the profitability coefficient (PC), the ratio of private and social profits or PC = (A - B - C)/(E - F - G), or D/H. The PC measures the incentive effects of all policies and thus serves as a proxy for the net policy transfer, since, L = (D - H). Its usefulness is restricted when private or social profits are negative, since, the signs of both entries must be known to allow clear interpretation.

A final incentive indicator is the subsidy ratio to producers (SRP), the net policy transfer as a proportion of total social revenues or SRP = L/E = (D - H)/E. The SRP shows the proportion of revenues in world prices that would be required if a single subsidy or tax were substituted for the entire set of commodity and macroeconomic policies. The SRP permits comparisons of the extent to which all policy subsidizes agricultural systems. The SRP measure can also be disaggregated into component transfers to show separately the effects of output, input, and factor policies.

DYNAMIC COMPARATIVE ADVANTAGE

The ability of an agricultural system to compete without distorting government policies can be strengthened or eroded by changes in economic conditions. Dynamic comparative advantage refers to shifts in a system's competitiveness that occur over time because of changes in three categories of economic parameters - long-run world prices of tradable outputs and inputs, social opportunity costs of domestic factors of production (labour, capital, and land), and production technologies used in farming or marketing. Together, these three parameters determine social profitability and comparative advantage.

The appropriate world prices for measuring efficiency or comparative advantage are long-run equilibrium levels that approximate best guesses of expected future prices. If the country's decisions to buy or sell on world markets will not have any measurable effect on world price levels, those price levels can be considered exogenous and, once arrived at, can be taken as given for domestic agricultural systems.

The world prices are the correct indicators of social valuation of tradable commodities even if a country's decisions to buy or sell internationally do affect the world price of a good. When a large country has market power, however, the analyst needs to take into account the impact of that country's trading decisions on world prices.

In the absence of knowledge of future prices, most analysts project constant long-run real prices rather than fluctuating prices. If new information results in changes in the constant price guess or in the projection of continually increasing or decreasing future prices, these changes can be incorporated easily into the PAM. Separate PAMs can be constructed for each year, and each can have different assumed world prices.

Costs of factor services in any country can be expected to change over time. But cyclical variations in the real wage and the real return to capital, associated with swings in macroeconomic policy, are not the primary focus of the PAM method. Instead, interest centers on long-run trends in the costs of labour, capital, and land.

As economies grow, real wages typically rise, both in absolute terms and relative to real costs of capital and land. For agricultural systems, changes in the social opportunity costs of labour and of capital depend on changes in the national environment for investment and growth. Land rental rates are endogenous to agriculture but will be constrained by changes in world prices and in real wage and interest rates, because payments to land and other permanently fixed factors come out of profits. Analysis of projected comparative advantage therefore includes both the future pressures that changing real factor prices might exert on agricultural systems and the influences of likely world prices for tradable outputs and inputs. The results identify systems that can readily expand and those that will have to contract or change in order to survive.

Changes over time in factor and commodity prices can also influence agricultural technologies. Farmers and researchers innovate, often by finding new ways of using less of factors that are relatively expensive (usually labour) and more of other inputs. Successful technological change permits commodities to be produced with reduced costs of one or more inputs. Empirical analysis of intra-system change can be done with partial budgeting, a technique in which individual cost-saving or revenue-increasing changes can be analyzed within the PAM for the initial system.

CONCLUDING COMMENTS

The central purpose of PAM analysis is to measure the impact of government policy on the private profitability of agricultural systems and on the efficiency of resource use. Private profitability and competitiveness are likely to be uppermost in the minds of those concerned specifically with agricultural incomes. Social profitability and efficiency are often emphasized by economic planners whose concern is the allocation of resources among sectors and the growth of aggregate income in the economy. Both sets of issues ultimately focus on the incentive effects of policy-part of the difference between private and social profitability-and on how policy incentives might be altered. Through evaluation of private and social revenues and costs, the PAM method is designed to illuminate these related issues of agricultural policy analysis. The approach is particularly well suited to empirical analysis of agricultural price policy and farm incomes, public investment policy and efficiency, and agricultural research policy and technological change.

The PAM approach to policy evaluation advocates a disaggregated view of efficiency effects (as measured by social profitability) and of non-efficiency effects. The analyst can do much in describing the contributions of a particular system to non-efficiency objectives and in quantifying implications for efficiency (aggregate income gains or losses). But it is left to the discretion of each policy-maker to determine whether tradeoffs between efficiency and non-efficiency objectives merit changes in policy or maintenance of incentives to particular systems.

In other approaches to policy analysis, it is desirable to aggregate measures of efficiency and non-efficiency effects into a single measure. Income distribution concerns, for example, can be introduced into the social cost estimates by weighting (with a value less than 1) of the efficiency-determined value of unskilled labour wages. Concerns for food self-sufficiency can be introduced by the addition of a premium to the world market value of output. If these weights were incorporated in the calculations of social profitability, the policy-makers' decision would become automatic and predetermined: encourage all systems with positive social profitability and discourage all systems with negative profitability.

The disadvantage of the aggregate approach lies in its tendency to lump together high-quality information (observable data on prices and input-output relationships) with relatively poor-quality information (implicit weights of society or of policy-makers regarding various prices of inputs and outputs). Moreover, attempts to quantify implicit policy weights presume the existence of some dictatorial policy-maker who speaks on behalf of society. Policy-making rarely occurs in such an environment. Policies are the outcomes of negotiated conflict between interest groups both within and outside the government. Quantitative studies provide improved information and thus increase the probability of good policy decisions. But these decisions, and the tradeoffs implied between efficiency and non-efficiency objectives, are the outcomes of debate based on this information, not inputs into the collection of information.

NEW AGRICULTURAL MARKETING POLICY

Agricultural Marketing Policy 2013 for "bringing in competition, enhancing capacity building, and ushering in an era of single unified licence market" in the State.

SIMPLE PROCEDURE

The licensing procedures would be simplified and a single licence would be made applicable for participants.

The policy envisages elimination of barriers to participation in markets to foster competition and ensure efficient determination of price, linking the primary market in the State to the national market for the benefit of all stakeholders in the marketing chain.

A panel, headed by Manoj R., Additional Secretary of the Cooperation Department, submitted the draft policy to the department in May 2013 after holding consultations with Indian Institute of Management Bangaluru and University of Agricultural Sciences.

The policy recommends warehousing-based sales, rationalisation of private markets, market fee waiver for perishable produce, and focus on direct purchase centres. The conditions that restrict participation would be removed to increase competition in the auction of agricultural produce.

It would reduce the role of middlemen and also help arrest unfair trade practices.

The policy aimed at adoption of technology required for setting up a comprehensive electronic auction system by regulated markets operating across the State for transparent price determination.

VIRTUAL MARKET

"A State-wide networked virtual market would be established by linking various regulated markets and warehouses, provided with assaying and grading facilities and other infrastructure. Farmers and other participants would have the choice to offer and sell in any regulated or private market in the State.

An enabling environment would be created to facilitate farmers to avail themselves of pledge loans to avoid distress sale.

Processes would be simplified and farmers would get online timely payment into their bank account.

The policy would enable the farmer to decide when to sell the produce and at what price, with a right to reject the price offered Karnataka Agricultural Produce Marketing (Regulation and Development) Act 1966 would be reviewed to facilitate the policy objectives and initiatives and to create a distinct, level-playing regulatory environment for transparent and efficient functioning of agricultural markets in the State.

GROWTH OF MARKETING INFRASTRUCTURE

Marketing infrastructure is most important not only for the performance of various marketing functions and for expansion of the size of market but also for transfer of appropriate price signals leading to improved marketing efficiency.

The projection production and marketable surplus of farm produce show that the volume of the commodities to be handled will be quite large. The paddy output available for milling will be around 155 million tonnes. The marketed surplus of cereals will be 102.74 million tonnes and pulses will be 15.20 million tonnes. Row cotton output is projected to be 3.2 million tonnes. The marketable surplus of perishable (fruits vegetables and livestock products) is anticipated to go up for present 172.69 million tonnes by 2006-07. The capacity to clean grade process store transport etc., to have to expand correspondingly to handle the additional marketed quantities.

PRESENT STATUS OF AGRICULTURAL MARKETING

Government organized marketing of agriculture in the country through the network of regulated markets established under the provisions of the Agricultural Produce Market Act enacted by the states and union territories. As on 31/3/2001 the markets covered under regulation is 7177. In addition there are 27924 rural periodical markets or *hats*. About 15 per cent of these in markets have been brought under the ambit of the regulation.

The regulated markets have helped in mitigating the market handicaps of producer's sellers. These have also provided physical facilities and institutional environment to the wholesalers commission agents. Traders and other functionaries for conducting activities.

It was envisaged that these regulated markets will provide facilities and services which would attract the farmers and buyers creating competitive trade environment thereby offering best of prices to the producer-sellers.

Studies of regulated markets show that they have achieved limited success in providing need based facilities and services conducive to achieving greater marketing efficiency. Most of these markets lack requisite facilities for handling the produce arriving in the yard. Rural markets in general and tribal hats in particular remained out of the ambit of the development.

Over a period of time these markets have acquired the status of institution with control and restrictions providing no help in direct marketing organised marketing organised retailing smooth supply of raw material to agro-processing units competitive trading information exchange adoption of innovative marketing system and technologies etc., as was envisaged under the provision of the Act.

Monolopolistic tendencies and practices have prevented development of free and competitive trade in primary markets future markets (or secondary markets) use of new tools and techniques in pre harvest management and post harvest management in handling exports agro based industries ware housing etc. The prominent activities like grading standardization scientific storage linked with finance search for suitable markets for excess of marketable surplus education of farmers in pre-and post-harvest management and facilities in the markets have become secondary activities. Marketing development funds have been siphoned to public ledger account by the state authorities adversely effecting modernization and infrastructure development vital for operational efficiency.

The worldwide Governments have recognized the importance of liberalising agricultural markets. In South Africa Agricultural Marketing is changed from controlled marketing to a free system. In Holland the growers' co-operatives are acquiring new companies a specialized in commodity export and import to achieve high degree of professionalism in marketing.

The ever increasing production spread of latest technologies changing socio economic environment increasing demand for downsizing the distribution chain reducing the marketing margins between the producers and the ultimate consumers challenges emerging out of liberalization and globalization in the post-WTO period required a vibrant dynamic and assimilative marketing structure and system.

REFORMS REQUIRED

In order to have vibrant competitive marketing systems the Government had to bring about reforms in existing policies rules and regulations with a view to remove all legal provisions inhibiting free marketing system. This is necessary to explore market access opportunities provided by liberalization.

Some of the legal provisions inhibiting development of vibrant dynamic and assimilative marketing structure and systems is the Agricultural Produce Marketing (Regulation) Act.

The Expert Committee on stregthnining and Development of Agricultural Marketing suggested various reforms in the statutory arrangements relating to agricultural marketing as well as policies and programmes for development and strengthening of agricultural marketing with a specific reference to needed investment package of incentives easy and adequate marketing credit.

This committee also recommended a review of the existing legal framework removal of restrictive provisions to promote competitive marketing structure to promote direct marketing by the farmers to improve price realisation to encourage forward and future trading to reduce price risk to induce increase the flow of funds to agricultural sector to support pledge financing by treating it as a direct priority sector lending and to promote market led extension to use of information technology for improving marketing services to the farmers.

A inter-ministerial task-force under the Chairmanship of Additional Secretary Department of Agriculture and Co-Operation Government. of India deliberated on these recommendations and identified nine areas. To work out action plan nine interministerial sub groups were constituted. The sub groups on legal reforms in Agricultural Marketing under the Chairmanship of the Joint Secretary (Marketing). The specific areas identified for reforms in the State Agricultural Produce Marketing Regulations Acts (APMC) are :-

Promotion of Integrated Markets in Private /Co-Operative Sector

Under the existing Act it is the state Governtment who are alone empowered to initiate the process of setting up a market for certain commodities to be regulated for a defined area, in which regulation is to be enforced under the provision of APMC Act. As a result of this provision the process of initiation of a market the service providers of agricultural marketing do not have any role. Thus, the service providers and/or any other individual or body of individuals cannot take initiatives for evaluation of viability and feasibility for setting up of a well-developed market with amenities and facilities required at a competitive cost.

A ball has been set rolling in Karnataka by providing additional chapter (XIII-A) in the Karnataka Agricultural Produce Marketing (Regulation) Act, 1966 to provide for the establishment of ' National Integrated Produce Market' owned and managed by the NDDB for the marketing of fruits vegetables and flowers. It is high time that all other States follow suit by amending the APMC Act in their respective states for providing establishment of alternate marketing structures in the shape of 'National Integrated Produce Market' to be owned and managed in the private sector or co-operatives sectors or farmers self help groups farmers associations private entrepreneurs or joint ventures. The service provider may be allowed to levy and collect service charge from the users who may be the producers - sellers or from the market users .

The integrated market infrastructure services will in addition to the physical infrastructure include: -

- Assembling
- Cleaning Sorting Grading packaging and quality certification
- Storage and finance
- Transport
- Retailing and Wholesaling
- E- trading
- Warehousing and pledge financing
- Value addition and
- Market information exchange service.

Direct Marketing

The direct marketing enables farmers to meet the specific demands of wholesalers or traders from the farmers inventory of graded and certified produce on one hand and of consumers based on consumers preference on the other hand helps the farmers to dynamically take advantage of favourable prices reduce marketing cost and thus their net margins. This encourages farmers to under take cleaning sorting grading and quality marking at the farm gate.

This will obviate the need to haul the produce to the regulated markets which are not necessarily equipped with all required services and facilities affecting the marketing efficiency adversely. It is reported that the consumers prices declined by the 20 to 30 per cent and producers received the prices rose by 10 to 20 per cent in South Korea as a consequence of expansion of direct marketing of Agricultural Products This model has been experimented in Punjab and Haryana (APNI MANDIS) Andhra Pradesh (Rythu Bazar) and Tamil Nadu (Uzhavar Santhaigal) . All the provision exists in various states market regulation act for direct marketing. Conserted efforts have not been made to promote the direct sales by the farmers to consumers or retailers without involving any intermediary in between . In a country such as ours there are large numbers of places where such markets could come up in organised sectors with private investment and can be developed in tuned with the time for forward and backward linkages.

Contract Farming

Contract Farming may be defined as an agreement between processing and/or marketing firms for production support at production support at predetermined prices. This stipulates a commitment on the part of the farmers to provide a specific commodity in terms of quality and quantity as determined by the purchaser and commitment on the part of company to support the farmer for production through inputs and other technical support contract farming is becoming popular in recent years and there are number of success stories like Maul NDDB PEPSI Co. etc. The contract farming needs to be further developed after identifying areas commodities and markets for market oriented and demand driven production planning. However while providing for this system of alternate marketing under the APMC Act it is necessary to draft any appropriate legislation separately for ensuring definition of terms and conditions of the agreement keeping in view the objectives.

Direct Contract between Producers and Processing Factories

Presently the farmers are not in position to enter into direct contract with the processors/ manufacturers located outside the market area as the commodity has to channels through regulated markets. Or in other words producer is not free to sale his produce by entering in to direct contract without attracting the provisions of this act whether inside or outside the market area. The direct contract between the producers and processing factories or bulk processors will provide monitory gains to the producers through improved competitiveness and the consumers by way of reasonable prices. The provision has to be made through an amendment in the present market act.

Direct Purchase from Farmers without any Licence

As per the provisions of the present Act once a market area is notified no person can set up establish or use any place for the purchase sale storage and regiment curing pressing or processing of any Agricultural produce or products of livestock or for the purchase or sale of livestock except in accordance with the condition of a license granted by the market committee. This provision prohibits free sale and purchase of agril. Commodities thereby adversely affecting the compititive pricing adds to marketing costs to the produces in the event of bringing the produce to the market yard for improving competitiveness and greater marketing efficiency and pricing edge the producer nay be allowed to sell the produce at the farm gate or threshing floor. This will promote direct contract between the producers and the processing factories with monitory gains to the producers processors and finally to the consumers in the free market.

Notification of Commodities

Under the APM Act the state government is empowered to notify agricultural commodities under the provisions of the Act for the purpose of regulation of marketing. This has resulted into anamoly. Even commodities not passing through the market yard or for which no services are provided are also notified. It is logical that only such commodities be notified which pass through the market yard or for which marketing infrastructure has been provided and the market fee should be in proportion to the infrastructure and facilities provided.

Single Point Levy of Market Fee

At present by and large market fee is collected on a particular lot whenever it is transacted. These amounts to multiple point collection Simplification of market fee and thus adding to the cost therefore it is necessary to introduce single point levy of market fee in the entire process of marketing in the country.

Tax

There is considerable variation in the structure of taxes and fees on the agricultural produce in various states. This distorts the operation of the domestic market giving wrong signals to the farmers adversely affecting the operational efficiency of the private trade farmer's co-operatives and public sector agencies. It is necessary to bring uniformity in the state level tax structure in agricultural commodities for improving the marketing efficiencies. As stated earlier in introductory remarks the ever-increasing marketable surplus will require matching infrastructure and facilities more so in the light of globalisation and liberalisation. The Working Group has estimated the potential investment of ₹ 11000 crores during the span of next 10 years a major part of which has to come from the private sector. For attracting the private sector investment of this order it is necessary to have proper orientation. This will require:

- Reducing the regulatory controls and simplifying the procedures;
- Active stance by the Central Government in some initiatives;
- Making complementary investment by the State Government and Central Government;
- Subsidizing a few activities to enable the private sector initiatives to attain viability; and
- Ensuring adequate credit flow to agricultural marketing activities.

AGRICULTURE AND ECONOMIC DEVELOPMENT

As a country develops economically, the relative importance of agriculture declines. The primary reason for this was shown by the 19th-century German statistician Ernst Engel, who discovered that as incomes increase the proportion of income spent on food declines. For example, if a family's income were to increase by 100 percent, the amount it would spend on food might increase by 60 percent; if formerly its expenditures on food had been 50 percent of its budget, after the increase they would amount to only 40 percent of its budget. It follows from this that, as incomes increase, a smaller fraction of the total resources of society is required to produce the amount of food demanded by the population.

PROGRESS IN FARMING

This fact would have surprised most economists of the early 19th century, who feared that the limited supply of land in the populated areas of Europe would determine that continent's ability to feed its growing population. Their fear was based on the so-called law of diminishing returns: that under given conditions an increase in the amount of labour and capital applied to a fixed amount of land results in a less than proportional increase in the output of food. This principle is a valid one, but what the classical economists could not foresee was the extent to which the state of the arts and the methods of production would change. Some of the changes occurred in agriculture; others occurred in other sectors of the economybut had a major effect on the supply of food.

In looking back upon the history of the more developed countries, one can see that agriculture has played an important part in the process of their enrichment. For one thing, if development is to occur, agriculture must be able to produce a surplus of food to maintain the growing non-agricultural labour force. Since, food is more essential for life than are the services provided by merchants or bankers or factories, an economy cannot shift to such activities unless food is available for barter or sale in sufficient quantities to support those engaged in them. Unless food can be obtained through international trade, a country does not normally develop industrially until its farm areas can supply its towns with food in exchange for the products of their factories.

Economic development also requires a growing labour force. In an agricultural country most of the workers needed must come from the rural population. Thus agriculture must not only supply a surplus of food for the towns, but it must also be able to produce the increased amount of food with a relatively smaller labour force. It may do so by substituting animal power for human power or by gradually introducing labour-saving machinery.

Agriculture may also be a source of the capital needed for industrial development to the extent that it provides a surplus that may be converted into the funds needed to purchase industrial equipment or to build roads and provide public services.

For these reasons a country seeking to develop its economy may be well advised to give a significant priority to agriculture. Experience in the developing countries has shown that agriculture can be made much more productive with the proper investment in irrigation systems, research, fertilizers, insecticides, and herbicides.

Fortunately, many advances in applied science do not require massive amounts of capital, although it may be necessary to expand marketing and transportation facilities so that farm output can be brought to the entire population.

One difficulty in giving priority to agriculture is that most of the increase in farm output and most of the income gains are concentrated in certain regions rather than extending throughout the country. The remaining farmers are not able to produce more and actually suffer a disadvantage as farm prices decline. There is no easy answer to this problem, but developing countries need to be aware of it; economic progress is consistent with lingering backwardness, as can be seen in parts of southern Italy or in the Appalachian area of the United States.

ECONOMIC DEVELOPMENT

Economic development, the process whereby simple, low-income national economies are transformed into modern industrial economies. Although the term is sometimes used as a synonym for economic growth, generally it is employed to describe a change in a country's economy involving qualitative as well as quantitative improvements. The theory of economic development—

how primitive and poor economies can evolve into sophisticated and relatively prosperous ones—is of critical importance to underdeveloped countries, and it is usually in this context that the issues of economic development are discussed.

Economic development first became a major concern after World War II. As the era of European colonialism ended, many former colonies and other countries with low living standards came to be termed underdeveloped countries, to contrast their economies with those of the developed countries, which were understood to be Canada, the United States, those of western Europe, most eastern European countries, the then Soviet Union, Japan, South Africa, Australia, and New Zealand. As living standards in most poor countries began to rise in subsequent decades, they were renamed the developing countries.

There is no universally accepted definition of what a developing country is; neither is there one of what constitutes the process of economic development. Developing countries are usually categorized by a per capita income criterion, and economic development is usually thought to occur as per capita incomes rise. A country's per capita income (which is almost synonymous with per capita output) is the best available measure of the value of the goods and services available, per person, to the society per year. Although there are a number of problems of measurement of both the level of per capita income and its rate of growth, these two indicators are the best available to provide estimates of the level of economic well-being within a country and of its economic growth.

It is well to consider some of the statistical and conceptual difficulties of using the conventional criterion of underdevelopment before analysing the causes of underdevelopment. The statistical difficulties are well known. To begin with, there are the awkward borderline cases. Even if analysis is confined to the underdeveloped and developing countries in Asia, Africa, and Latin America, there are rich oil countries that have per capita incomes well above the rest but that are otherwise underdeveloped in their general economic characteristics. Second, there are a number of technical difficulties that make the per capita incomes of many underdeveloped countries (expressed in terms of an international currency, such as the U.S., dollar) a very crude measure of their per capita real income. These difficulties include the defectiveness of the basic national income and population statistics, the inappropriateness of the official exchange rates at which the national incomes in terms of the respective domestic currencies are converted into the common denominator of the U.S., dollar, and the problems of estimating the value of the non-cash components of real incomes in the underdeveloped countries. Finally, there are conceptual problems in interpreting the meaning of the international differences in the per capita income levels. Although the difficulties with income measures are well established, measures of per capita income correlate reasonably well with other measures of economic well-being, such as life expectancy, infant mortality rates, and literacy rates. Other indicators, such as nutritional status and the per capita availability of hospital beds, physicians, and teachers, are also closely related to per capita income levels. While a difference of, say, 10 percent in per capita incomes between two countries would not be regarded as necessarily indicative of a difference in living standards between them, actual observed differences are of a much larger magnitude. India's per capita income, for example, was estimated at $270 in 1985. In contrast, Brazil's was estimated to be $1,640, and Italy's was $6,520. While economists have cited a number of reasons why the implication that Italy's living standard was 24 times greater than India's might be biased upward, no one would doubt that the Italian living standard was significantly higher than that of Brazil, which in turn was higher than India's by a wide margin.

The interpretation of a low per capita income level as an index of poverty in a material sense may be accepted with two qualifications. First, the level of material living depends not on per capita income as such but on per capita consumption.

The two may differ considerably when a large proportion of the national income is diverted from consumption to other purposes; for example, through a policy of forced saving. Second, the poverty of a country is more faithfully reflected by the representative standard of living of the great mass of its people. This may be well below the simple arithmetic average of per capita income or consumption when national income is very unequally distributed and there is a wide gap in the standard of living between the rich and the poor.

The usual definition of a developing country is that adopted by the World Bank: "low-income developing countries" in 1985 were defined as those with per capita incomes below $400; "middle-income developing countries" were defined as those with per capita incomes between $400 and $4,000. To be sure, countries with the same per capita income may not otherwise resemble one another: some countries may derive much of their incomes from capital-intensive enterprises, such as the extraction of oil, whereas other countries with similar per capita incomes may have more numerous and more productive uses of their labour force to compensate for the absence of wealth in resources. Kuwait, for example, was estimated to have a per capita income of $14,480 in 1985, but 50 percent of that income originated from oil. In most regards, Kuwait's economic and social indicators fell well below what other countries with similar per capita incomes had achieved. Centrally planned economies are also generally regarded as a separate class, although China and North Korea are universally considered developing countries. A major difficulty is that prices serve less as indicators of relative scarcity in centrally planned economies and hence, are less reliable as indicators of the per capita availability of goods and services than in market-oriented economies.

Estimates of percentage increases in real per capita income are subject to a somewhat smaller margin of error than are estimates of income levels. While year-to-year changes in per capita income are heavily influenced by such factors as weather (which affects agricultural output, a large component of income in most developing countries), a country's terms of trade, and other factors, growth rates of per capita income over periods of a decade or more are strongly indicative of the rate at which average economic well-being has increased in a country.

ECONOMIC DEVELOPMENT AS AN OBJECTIVE OF POLICY

Motives for Development

The field of development economics is concerned with the causes of underdevelopment and with policies that may accelerate the rate of growth of per capita income. While these two concerns are related to each other, it is possible to devise policies that are likely to accelerate growth (through, for example, an analysis of the experiences of other developing countries) without fully understanding the causes of underdevelopment.

Studies of both the causes of underdevelopment and of policies and actions that may accelerate development are undertaken for a variety of reasons. There are those who are concerned with the developing countries on humanitarian grounds; that is, with the problem of helping the people of these countries to attain certain minimum material standards of living in terms of such factors as food, clothing, shelter, and nutrition. For them, low per capita income is the measure of the problem of poverty in a material sense. The aim of economic development is to improve the material standards of living by raising the absolute level of per capita incomes. Raising per capita incomes is also a stated objective of policy of the governments of all developing countries. For policy-makers and economists attempting to achieve their governments' objectives, therefore, an understanding of economic development, especially in its policy dimensions, is important. Finally, there are those who are concerned with economic development either because they believe it is what people in developing countries want or because they believe that political stability can be assured only with satisfactory rates of economic growth. These motives are not mutually exclusive. Since, World War II many industrial countries have extended foreign aid to developing countries for a combination of humanitarian and political reasons.

Those who are concerned with political stability tend to see the low per capita incomes of the developing countries in relative terms; that is, in relation to the high per capita incomes of the developed countries. For them, even if a developing country is able to improve its material standards of living through a rise in the level of its per capita income, it may still be faced with the more intractable subjective problem of the discontent created by the widening gap in the relative levels between itself and the richer countries. (This effect arises simply from the operation of the arithmetic of growth on the large initial gap between the income levels of the developed and the underdeveloped countries. As an example, an underdeveloped country with a per capita income of $100 and a developed country with a per capita income of $1,000 may be considered. The initial gap in their incomes is $900. Let the incomes in both countries grow at 5 percent. After one year, the income of the underdeveloped country is $105, and the income of the developed country is $1,050. The gap has widened to $945. The income of the underdeveloped country would have to grow by 50 percent to maintain the same absolute gap of $900.) Although there was once in development economics a debate as to whether raising living standards or reducing the relative gap in living standards was the true desideratum of policy, experience during the 1960–80 period convinced most observers that developing countries could, with appropriate policies, achieve sufficiently high rates of growth both to raise their living standards fairly rapidly and to begin closing the gap.

IDENTIFYING PROJECT COSTS AND BENEFITS

We undertake economic analyses of agricultural projects to compare costs with benefits and determine which among alternative projects have an acceptable return. The costs and benefits of a proposed project therefore must be identified. Furthermore, once costs and benefits are known, they must be priced, and their economic values determined. All of this is obvious enough, but frequently it is tricky business.

Objectives, Costs, and Benefits

In project analysis, the objectives of the analysis provide the standard against which costs and benefits are defined. Simply put, a cost is anything that reduces an objective, and a benefit is anything that contributes to an objective.

The problem with such simplicity, however, is that each participant in a project has many objectives. For a farmer, a major objective of participating is to maximize the amount his family has to live on. But this is only one of the farmer's interests. He may also want his children to be educated; as a result, they may not be available to work full time in the fields. He may also value his time away from the fields: a farmer will not adopt a cropping pattern, however remunerative, that requires him to work ten hours a day 365 days a year. Taste preference may lead a farmer to continue to grow a traditional variety of rice for home consumption even though a new, high-yielding variety might increase his family income more. A farmer may wish to avoid risk, and so may plan his cropping pattern to limit the risk of crop failure to an acceptable level or to reduce the risk of his depending solely on the market for the food grains his family will consume. As a result, although he may be able to increase his income over time if he grows cotton instead of wheat or maize, he would rather continue growing food grains to forestall the possibility that in any one year the cotton crop might fail or that food grains might be available for purchase in the market only at a very high price. All these considerations affect a farmer's choice of cropping pattern and thus the income-generating capacity of the project. Yet all are sensible decisions in the farmer's view. In the analytical system presented here, we will try to identify the cropping pattern that we think the farmer will most probably select, and then we will judge the effects of that pattern on his incremental income and, thus, on the new income generated by the project.

For private business firms or government corporations, a major objective is to maximize net income, yet both have significant objectives other than simply making the highest profit possible. Both will want to diversify their activities to reduce risk. The private store owner may have a preference for leisure, which leads him to hire a manager to help operate his store, especially during late hours. This reduces the income-since, the manager must be paid a salary-but it is a sensible choice. For policy reasons, a public bus corporation may decide to maintain services even in less densely populated areas or at off-peak hours and thereby reduce its net income. In the analytical system here, we first identify the operating pattern that the firms in the project will most likely follow and then build the accounts to assess the effects of that pattern on the income-generating capacity of the project.

A society as a whole will have as a major objective increased national income, but it clearly will have many significant, additional objectives. One of the most important of these is income distribution. Another is simply to increase the number of productive job opportunities so that unemployment may be reduced-which may be different from the objective of income distribution itself. Yet another objective may be to increase the proportion of savings in any given period so there will be more to Invest, faster growth, and, hence, more income in the future. Or, there may be issues to address broader than narrow economic considerations-such as the desire to increase regional integration, to upgrade the general level of education, to improve rural health, or to safeguard national security. Any of these objectives might lead to the choice of a project (or a form of a project) that is not the alternative that would contribute most to national income narrowly defined.

No formal analytical system for project analysis could possibly take into account all the various objectives of every participant in a project. Some selection will have to be made. In the analytical system here, we will take as formal criteria very straightforward objectives of income maximization and accommodate other objectives at other points in the process of project selection. The justification for this is that in most developing countries increased income is probably the single most important objective of individual economic effort, and increased national income is probably the most important objective of national economic policy.

For farms, we will take as the objective maximizing the incremental net benefit-the increased amount the farm family has to live on as a result of participating in the project. For a private business firm or corporation in the public sector, we will take as the objective maximizing the incremental net income. And for the economic analysis conducted from the standpoint of the society as a whole, we will take as the objective maximizing the contribution the project makes to the national income-the value of all final goods and services produced during a particular period, generally a year. This is virtually the same objective, except for minor formal variations in definition, as maximizing gross domestic product (GDP). It is important to emphasize that taking the income a project will contribute to a society as the formal analytical criterion in economic, analysis does not downgrade other objectives or preclude our considering them. Rather, we will simply treat consideration of other objectives as separate decisions. Using our analytical system, we can judge which among alternative projects or alternative forms of a particular project will make an acceptable contribution to national income. This will enable us to recommend to those who must make the Investment decision a project that has a high income-generating potential and also will make a significant contribution to other social objectives. For example, from among those projects that make generally the same contribution to increased income, we can choose the one that has the most favourable effects on income distribution, or the one that creates the most jobs, or the one that is the most attractive among those in a disadvantaged region.

Thus, in the system of economic analysis, anything that reduces national income is a cost and anything that increases national income is a benefit. Since, our objective is to increase the sum of all final goods and services, anything that directly reduces the total final goods and services is obviously a cost, and anything that directly increases them is clearly a benefit. But recall, also, the intricate workings of the economic system. When the project analyzed uses some intermediate good or service-something that is used to produce something else - by a chain of events it eventually reduces the total final goods and services available elsewhere in the economy. On the one hand, if we divert an orange that can be used for direct consumption-and thus is a final good-to the production of orange juice, also a final good, we are reducing the total available final goods and services, or national income, by the value of the orange and increasing it by the value of the orange juice. On the other hand, if we use cement to line an irrigation canal, we are not directly reducing the final goods and services available; instead, we are simply reducing the availability of an intermediate good. But the consequence of using the cement in the irrigation project is to shift the cement away from some other use in the economy. This, in turn, reduces production of some other good, and so on through the chain of events until, finally, the production of final goods and services, the national income, is reduced. Thus, using cement in the project is a cost to the economy. How much the national income will be reduced by using the cement for the project is part of what we must estimate when we turn, to deriving economic values. On the benefit side, we have a similar pattern. Lining a canal increases available water that, in turn, may increase wheat production, and so on through a chain of events until in the end the total amount of bread is increased.

By this mechanism, the project leads to an increase in the total amount of final goods and services, which is to say it increases the national income. Again, part of the analyst's task in the economic analysis is to estimate the amount of this increase in national income available to the society; that is, to determine whether, and by how much, the benefits exceed the costs in terms of national income. If this rather simple definition of economic costs and benefits is kept in mind, possible confusion will be avoided when shadow prices are used to value resource flows.

Note that, by defining our objective for economic analysis in terms of change in national income, we are defining it in real terms. (Real terms, as opposed to money terms, refer to the physical, tangible characteristics of goods and services.) To an important degree, economic analysis, in contrast to financial analysis, consists in tracing the real resource flows induced by an Investment rather than the Investment's monetary effects.

With these objectives defined, we may then say that in financial analysis our numeraire-the common measurement used as the unit of account-is a unit of currency, generally domestic currency, whereas in economic analysis our numeraire is a unit of national income, generally also expressed in domestic currency.

In the economic analysis we will assume that all financing for a project comes from domestic sources and that all returns from the Project Go to domestic residents. [This is one reason why we identify our social objective with the gross domestic product (GDP) instead of the more familiar gross national product (GNP).] This convention-almost universally accepted by project analysts-separates the decision of how good a project is in its income-generating potential from the decision of how to finance it. The actual terms of financing available for a particular project will not influence the evaluation. Instead, we will assume that the proposed project is the best Investment possible and that financing will then be sought for it at the best terms obtainable. This convention serves well whenever financing can be used for a range of projects or even versions of roughly the same project. The only case in which it does not hold well is the rather extreme case in which foreign financing is very narrowly tied to a particular project and will be lost if the project is not implemented. Then the analyst may be faced with the decision of implementing a lower-yielding project with foreign financing or choosing a higher-yielding alternative but losing the foreign loan.

"With" and "Without" Comparisons

Project analysis tries to identify and value the costs and benefits that will arise with the proposed project and to compare them with the situation as it would be without the project. The difference is the incremental net benefit arising from the project investment. This approach is not the same as comparing the situation "before" and "after" the project. The before-and-after comparison fails to account for changes in production that would occur without the project and thus leads to an erroneous statement of the benefit attributable to the project investment.

A change in output without the project can take place in two kinds of situations. The most common is when production in the area is already growing, if only slowly, and will probably continue to grow during the life of the project. The objective of the project is to increase growth by intensifying production. In Syria at the time the First Livestock Development Project was appraised, for example, production in the national sheep flock was projected to grow at about 1 percent a year without the project. The project was to increase and stabilize sheep production and the incomes of seminomadic flock owners and sheep fatteners by stabilising the availability of feed and improving veterinary services. With the project, national flock production was projected to grow at the rate of 3 percent a year. In this case, if the project analyst had simply compared the output before and after the project, he would have erroneously attributed the total increase in sheep production to the project investment. Actually, what can be attributed to the project investment is only the 2 percent incremental increase in production in excess of the 1 percent that would have occurred anyway.

A change in output can also occur without the project if production would actually fall in the absence of new Investment. In Guyana, on the north coast of South America, rice and sugarcane are produced on a strip of clay and silt soil edging the sea. The coast was subject to erosion from wave action. Under the Sea Defence Project, the government of Guyana has built seawalls to prevent the erosion. The benefit from this project, then, is not increased production but avoiding the loss of agricultural output and sites for housing. A simple before-and-after comparison would fail to identify this benefit.

In some cases, an Investment to avoid a loss might also lead to an increase in production, so that the total benefit would arise partly from the loss avoided and partly from increased production. In Pakistan, many areas are subject to progressive salinization as a result of heavy irrigation and the waterlogging that is in part attributable to seepage from irrigation canals. Capillary action brings the water to the surface where evaporation occurs, leaving the salt on the soil. If nothing is done to halt the process, crop production will fall. A project is proposed to line some of the canals, thus to reduce the seepage and permit better drainage between irrigations. The proposed project is expected to arrest salinization, to save for profitable use the irrigation water otherwise lost to seepage, and to help farmers increase their use of modern inputs. The combination of measures would not only avoid a loss but also lead to an increase in production. Again, a simple before-and-after comparison would fail to identify the benefit realized by avoiding the loss.

Of course, if no change in output is expected in the project area without the project, then the distinction between the before-and-after comparison and the with-and-without comparison is less crucial. In some projects the prospects for increasing production without new INVESTMENT are minimal. In the Kemubu Irrigation Project in northeastern Malaysia, a pump irrigation scheme was built that permitted farmers to produce a second rice crop during the dry season. Without the project, most of the area was used for grazing, and with the help of residual moisture or small pumps some was used to produce tobacco and other cash crops. Production was not likely to increase because of the limited amount of water available. With the project now in operation, rice is grown in the dry season. Of course, the value of the second rice crop could not be taken as the total benefit from the project. From this value must be deducted the value forgone from the grazing and the production of cash crops. Only the incremental value could be attributed to the new INVESTMENT in pumps and canals.

Another instance where there may be no change in output without the project is the obvious one found in some settlement projects. Without the project there may be no economic use of the area at all. In the Alto Turi Land Settlement Project in northeastern Brazil, settlers established their holdings by clearing the forest, planting upland rice, and then establishing pasture for production of beef cattle. At the time the settlers took up their holdings the forest had not been economically exploited-nor was it likely to be, at least for many years, in the absence of the project. In this case, the output without the project would be the same as the output before the project.

Direct Transfer Payments

Some entries in financial accounts really represent shifts in claims to goods and services from one entity in the society to another and do not reflect changes in national income. These are the so-called direct transfer payments, which are much easier to identify if our definition of costs and benefits is kept in mind. In agricultural project analysis four kinds of direct transfer payments are common: taxes, subsidies, loans, and debt service (the payment of interest and repayment of principal). Take taxes, for example. In financial analysis a tax payment is clearly a cost. When a farmer pays a tax, his net benefit is reduced.

But the farmer's payment of tax does not reduce the national income. Rather, it transfers income from the farmer to the government so that this income can be used for social purposes presumed to be more important to the society than the increased individual consumption (or investment) had the farmer retained the amount of the tax. Because payment of tax does not reduce national income, it is not a cost from the standpoint of the society as a whole. Thus, in economic analysis we would not treat the payment of taxes as a cost in project accounts. Taxes remain a part of the overall benefit stream of the project that contributes to the increase in national income.

Of course, no matter what form a tax takes, it is still a transfer payment-whether a direct tax on income or an indirect tax such as a sales tax, an excise tax, or a tariff or duty on an imported input for production. But some caution is advisable here. Taxes that are treated as a direct transfer payment are those representing a diversion of net benefit to the society. Quite often, however, government charges for goods supplied or services rendered may be called taxes. Water rates, for example, may be considered a tax by the farmer, but from the standpoint of the society as a whole they are a payment by the farmer to the irrigation authority in exchange for water supplied. Since, building the irrigation system reduces national income, the farmer's payment for the water is part of the cost of producing the crop, the same as any other payment for a production input. Other payments called taxes may also be payments for goods and services rendered rather than transfers to the government. A stevedoring charge at the port is not a tax but a payment for services and so would not be treated as a duty would be. Whether a tax should be treated as a transfer payment or as a payment for goods and services depends on whether the payment is a compensation for goods and services needed to carry out the project or merely a transfer, to be used for general social purposes, of some part of the benefit from the project to the society as a whole.

Subsidies are simply direct transfer payments that flow in the opposite direction from taxes. If a farmer is able to purchase fertilizer at a subsidized price, that will reduce his costs and thereby increase his net benefit, but the cost of the fertilizer in the use of the society's real resources remains the same. The resources needed to produce the fertilizer (or import it from abroad) reduce the national income available to the society. Hence, for economic analysis of a project we must enter the full cost of the fertilizer.

Again, it makes no difference what form the subsidy takes. One form is that which lowers the selling price of inputs below what otherwise would be their market price. But a subsidy can also operate to increase the amount the farmer receives for what he sells in the market, as in the case of a direct subsidy paid by the government that is added to what the farmer receives in the market. A more common means to achieve the same result does not involve direct subsidy. The market price may be maintained at a level higher than it otherwise would be by, say, levying an import duty on competing imports or forbidding competing imports altogether. Although it is not a direct subsidy, the difference between the higher controlled price set by such measures and the lower price for competing imports that would prevail without such measures does represent an indirect transfer from the consumer to the farmer.

Credit transactions are the other major form of direct transfer payment in agricultural projects. From the standpoint of the farmer, receipt of a loan increases the production resources he has available; payment of interest and repayment of principal reduce them. But from the stand-point of the economy, things look different. Does the loan reduce the national income available? No, it merely transfers the control over resources from the lender to the borrower. Perhaps one farmer makes the loan to his neighbour. The lending farmer cannot use the money he lends to buy fertilizer, but the borrowing farmer can. The use of the fertilizer, of course, is a cost to the society because it uses up resources and thus reduces the national income. But the loan transaction does not itself reduce the national income; it is, rather, a direct transfer payment. In reverse, the same thing happens when the farmer repays his loan. The farmer who borrowed cannot buy fertilizer with the money he uses to repay the loan his neighbour made, but his neighbour can. Thus, the repayment is also a direct transfer payment.

Some people find the concept of transfer payments easier to understand if it is stated in terms of real resource flows. Taking this approach in economic analysis, we see that a tax does not represent a real resource flow; it represents only the transfer of a claim to real resource flows. The same holds true for a direct subsidy that represents the transfer of a claim to real resources from, say, an urban consumer to a farmer. This line of reasoning also applies to credit transactions. A loan represents the transfer of a claim to real resources from the lender to the borrower. When the borrower pays interest or repays the principal, he is transferring the claim to the real resources back to the lender-but neither the loan nor the repayment represents, in itself, use of the resources.

Costs of Agricultural Projects

In almost all project analyses, costs are easier to identify (and value) than benefits. In every instance of examining costs, we will be asking ourselves if the item reduces the net benefit of a farm or the net income of a firm (our objectives in financial analysis), or the national income (our objective in economic analysis).

Physical Goods

Rarely will physical goods used in an agricultural project be difficult to identify. For such goods as concrete for irrigation canals, fertilizer and pesticides for increasing production, or materials for the construction of homes in land settlement projects, it is not the identification that is difficult but the technical problems in planning and design associated with finding out how much will be needed and when.

Labour

Neither will the labour component of agricultural projects be difficult to identify. From the highly skilled project manager to the farmer maintaining his orchard while it is coming into production, the labour inputs raise less a question of what than of how much and when. Labour may, however, raise special valuation problems that call for the use of a shadow price. Confusion may also arise on occasion in valuing family labour.

Land

By the same reasoning, the land to be used for an agricultural project will not be difficult to identify. It generally is not difficult to determine where the land necessary for the project will be located and how much will be used. Yet problems may arise in valuing land because of the very special kind of market conditions that exist when land is transferred from one owner to another.

Contingency Allowances

In projects that involve a significant initial INVESTMENT in civil works, the construction costs are generally estimated on the initial assumption that there will be no modifications in design that would necessitate changes in the physical work; no exceptional conditions such as unanticipated geological formations; and no adverse phenomena such as floods, landslides, or unusually bad weather. In general, project cost estimates also assume that there will be no relative changes in domestic or international prices and no inflation during the investment period. It would clearly be unrealistic to rest project cost estimates only on these assumptions of perfect knowledge and complete price stability. Sound project planning requires that provision be made in advance for possible adverse changes in physical conditions or prices that would add to the baseline costs. Contingency allowances are thus included as a regular part of the project cost estimates.

Contingency allowances may be divided into those that provide for physical contingencies and those for price contingencies. In turn, price contingency allowances comprise two categories, those for relative changes in price and those for general inflation. Physical contingencies and price contingencies that provide for increases in relative costs underlie our expectation that physical changes and relative price changes are likely to occur, even though we cannot forecast with confidence just how their influence will be felt. The increase in the use of real goods and services represented by the physical contingency allowance is a real cost and will reduce the final goods and services available for other purposes; that is, it will reduce the national income and, hence, is a cost to the society. Similarly, a rise in the relative cost of an item implies that its productivity elsewhere in the society has increased; that is, its potential contribution to national income has risen. A greater value is forgone by using the item for our project; hence, there is a larger reduction in national income. Physical contingency allowances and price contingency allowances for relative changes in price, then, are expected-if unallocated-project costs, and they properly form part of the cost base when measures of project worth are calculated. General inflation, however, poses a different problem. The future prices, in project analysis the most common means of dealing with inflation is to work in constant prices, on the assumption that all prices will be affected equally by any rise in the general price level. This permits valid comparisons among alternative projects. If inflation is expected to be significant, however, provision for its effects on project costs needs to be made in the project financing plan so that an adequate budget is obtained. Contingency allowances for inflation would not, however, be included among the costs in project accounts other than the financing plan.

Taxes

Recall that the payment of taxes, including duties and tariffs, is customarily treated as a cost in financial analysis but as a transfer payment in economic analysis (since, such payment does not reduce the national income). The amount that would be deducted for taxes in the financial accounts remains in the economic accounts as part of the incremental net benefit and, thus, part of the new income generated by the project.

Debt Service

The same approach applies to debt service-the payment of interest and the repayment of capital. Both are treated as an outflow in financial analysis. In economic analysis, however, they are considered transfer payments and are omitted from the economic accounts. Treatment of interest during construction can give rise to confusion. Lending institutions sometimes add the value of interest during construction to the principal of the loan and do not require any interest payment until the project begins to operate and its revenues are flowing. This process is known as "capitalising" interest. The amount added to the principal as a result of capitalising interest during construction is similar to an additional loan. Capitalising interest defers interest cost, but when the interest payments are actually due, they will, of course, be larger because the amount of the loan has been increased. From the standpoint of economic analysis, the treatment of interest during construction is clear. It is a direct transfer payment the same as any other interest payment, and it should be omitted from the economic accounts. Often interest during construction is simply added to the capital cost of the project. To obtain the economic value of the capital cost, the amount of the interest during construction must be subtracted from the capital cost and omitted from the economic account. In economic analysis, debt service is treated as a transfer within the economy even if the project will actually be financed by a foreign loan and debt service will be paid abroad. This is because of the convention of assuming that all financing for a project will come from domestic sources and all returns from the project will go to domestic residents. This convention, as noted earlier, separates the decision of how good a project is from the decision of how to finance it. Hence, even if it were expected that a project would be financed, say, by a World Bank loan, the debt service on that loan would not appear as a cost in the economic accounts of the project analysis.

Sunk Costs

Sunk costs are those costs incurred in the past upon which a proposed new investment will be based. Such costs cannot be avoided, however poorly advised they may have been. When we analyze a proposed investment, we consider only future returns to future costs; expenditures in the past, or sunk costs, do not appear in our accounts.

In practice, if a considerable amount has already been spent on a project, the future returns to the future costs of completing the project would probably be quite attractive even if it is clear in retrospect that the project should never have been begun. The ridiculous extreme is when only one dollar is needed to complete a project, even a rather poor one, and when no benefit can be realized until the project is completed. The "return" to that last dollar may well be extremely high, and it would be clearly worthwhile to spend it. But the argument that because much has already been spent on a project it therefore must be continued is not a valid criterion for decision. There are cases in which it would be preferable simply to stop a project midway or to draw it to an early conclusion so that future resources might be freed for higher-yielding alternatives.

For evaluating past Investment decisions, it is often desirable to do an economic and financial analysis of a completed project. Here, of course, the analyst would compare the return from all expenditures over the past life of the project with all returns. But this kind of analysis is useful only for determining the yield of past projects in the hope that judgements about future projects may be better informed. It does not help us decide what to do in the present. Money spent in the past is already gone; we do not have as one of our alternatives not to implement a completed project.

TANGIBLE BENEFITS OF AGRICULTURAL PROJECTS

Tangible benefits of agricultural projects can arise either from an increased value of production or from reduced costs. The specific forms in which tangible benefits appear, however, are not always obvious, and valuing them may be quite difficult.

Increased Production

Increased physical production is the most common benefit of agricultural projects. An irrigation project permits better water control so that farmers can obtain higher yields. Young trees are planted on cleared jungle land to increase the area devoted to growing oil palm. A credit project makes resources available for farmers to increase both their operating expenditures for current production-for fertilizers, seeds, or pesticides-and their INVESTMENT for a tubewell or a power thresher. The benefit is the increased production from the farm.

In a large proportion of agricultural projects the increased production will be marketed through commercial channels. In that case identifying the benefit and finding a market price will probably not prove too difficult, although there may be a problem in determining the correct value to use in the economic analysis.

In many agricultural projects, however, the benefits may well include increased production consumed by the farm family itself. Such is the case in irrigation rehabilitation projects along the north coast of Java. The home-consumed production from the projects increased the farm families' net benefit and the national income just as much as if it had been sold in the market. Indeed, we could think of the hypothetical case of a farmer selling his output and then buying it back.

Since, home-consumed production contributes to project objectives in the same way as marketed production, it is clearly part of the project benefits in both financial and economic analysis. Omitting home-consumed production will tend to make projects that produce commercial crops seem relatively high-yielding, and it could lead to a poor choice among alternative projects. Failure to include home-consumed production will also mean underestimating the return to agricultural Investments relative to investments in other sectors of the economy.

When home-consumed crops will figure prominently in a project, the importance of careful financial analysis is increased. In this case, it is necessary to estimate not only the incremental net benefit-including the value of home-consumed production and money from off-farm sales-but also the cash available to the farmer.

From the analysis of cash income and costs, one can determine if farmers will have the cash in hand to purchase modern inputs or to pay their credit obligations. It is possible to have a project in which home-consumed output increases enough for the return to the economy as a whole to be quite attractive, but in which so little of the increased production is sold that farmers will not have the cash to repay their loans.

Quality Improvement

In some instances, the benefit from an agricultural project may take the form of an improvement in the quality of the product. For example, the analysis for the Livestock Development Project in Ecuador, which was to extend loans to producers of beef cattle, assumed that ranchers would be able not only to increase their cattle production but also to improve the quality of their animals so that the average live price of steers per kilogram would rise from S/5.20 to S/6.40 in constant value terms over the twelve-year development period. (The symbol for Ecuadorian sucres is S/.) Loans to small dairy farmers in the Rajasthan Smallholder Dairy Improvement Project in India are intended to enable farmers not only to increase output but also to improve the quality of their product. Instead of selling their milk to make ghee (cooking oil from clarified butter), farmers will be able to sell it for a higher price in the Jaipur fluid milk market. As in these examples, both increased production and quality improvement are most often expected in agricultural projects, although both may not always be expected. One word of warning: both the rate and the extent of the benefit from quality improvement can easily be overestimated.

Change in Time of Sale

In some agricultural projects, benefits will arise from improved marketing facilities that allow the product to be sold at a time when prices are more favourable. A grain storage project may make it possible to hold grain from the harvest period, when the price is at its seasonal low, until later in the year when the price has risen. The benefit of the storage investment arises out of this change in "temporal value".

Change in Location of Sale

Other projects may include Investment in trucks and other transport equipment to carry products from the local area where prices are low to distant markets where prices are higher. For example, the Fruit and Vegetable Export Project in Turkey included provision for trucks and ferries to transport fresh produce from southeastern Turkey to outlets in the European Common Market. The benefits of such projects arise from the change in "locational value."

In most cases the increased value arising from marketing projects will be split between farmers and marketing firms as the forces of supply and demand increase the price at which the farmer can sell in the harvest season and reduce the monopolistic power of the marketing firm or agency. Many projects are structured to ensure that farmers receive a larger part of the benefit by making it possible for them to build storage facilities on their farms or to band together into cooperatives, but an agricultural project could also involve a private marketing firm or a government agency, in which case much of the benefit could accrue to someone other than farmers.

Changes in Product Form (Grading and Processing)

Projects involving agricultural processing industries expect benefits to arise from a change in the form of the agricultural product. Farmers sell paddy rice to millers who, in turn, sell polished rice. The benefit to the millers arises from the change in form. Canners preserve fruit, changing its form and making it possible at a lower cost to change its time or location of sale. Even a simple processing facility such as a grading shed gives rise to a benefit through changing the form of the product from run-of-the-orchard to sorted fruit. In the Himachal Pradesh Apple Marketing Project in northern India, the value of the apples farmers produce is increased by sorting; the best fruit is sold for fresh consumption while fruit of poorer quality is used to make a soft drink concentrate. In the process, the total value of the apples is increased.

Cost reduction through Mechanization

The classic example of a benefit arising from cost reduction in agricultural projects is that gained by Investment in agricultural machinery to reduce labour costs.

Examples are tubewells substituting for hand-drawn or animal-drawn water, pedal threshers replacing hand threshing, or (that favourite example) tractors replacing draft animals. Total production may not increase, but a benefit arises because the costs have been trimmed (provided, of course, that the gain is not offset by displaced labour that cannot be productively employed elsewhere).

Reduced Transport Costs

Cost reduction is a common source of benefit wherever transport is a factor. Better feeder roads or highways may reduce the cost of moving produce from the farm to the consumer. The benefit realized may be distributed among farmers, truckers, and consumers.

Losses Avoided

In discussing with-and-without comparisons in project analyses earlier in this chapter, we noted that in some projects the benefit may arise not from increased production but from a loss avoided. This kind of benefit stream is not always obvious, but it is one that the with-and-without test tends to point out clearly. In Jamaica, lethal yellowing is attacking the Jamaica Tall variety of coconut.

The government has undertaken a large Investment to enable farmers to plant Malayan Dwarf coconuts, which are resistant to the disease. Total production will change very little as a result of the investment, yet both the farmers and the economy will realize a real benefit because the new investment prevents loss of income. The Lower Egypt Drainage Project involves the largest single tile drainage system in the world. The benefit will arise not from increasing production in the already highly productive Nile delta, but from avoiding losses due to the waterlogging caused by year-round irrigation from the Aswan High Dam.

Sometimes a project increases output through avoiding loss-a kind of double classification, but one that in practice causes no problem. Proposals to eradicate foot-and-mouth disease in Latin America envision projects by which the poor physical condition or outright death of animals will be avoided. At the same time, of course, beef production would be increased.

Other Kinds of Tangible Benefits

Although we have touched on the most common kinds of benefits from agricultural projects, those concerned with agricultural development will find other kinds of tangible, direct benefits most often in sectors other than agriculture. Transport projects are often very important for agricultural development. Benefits may arise not only from cost reduction, as noted earlier, but also from time savings, accident reduction, or development activities in areas newly accessible to markets. If new housing for farmers has been included among the costs of a project, as is often the case in land settlement and irrigation projects, then among the benefits will be an allowance for the rental value of the housing. Since, this is an imputed value, there are valuation problems that will be noted later.

SECONDARY COSTS AND BENEFITS

Projects can lead to benefits created or costs incurred outside the project itself. Economic analysis must take account of these external, or secondary, costs and benefits so they can be properly attributed to the project Investment. (Of course, this applies only in economic analysis; the problem does not arise in financial analysis.)

When market prices are used in economic analysis, as has been the custom in the United States for water resource and other public works projects, it is necessary to estimate the secondary costs and benefits and then add them to the direct costs and benefits. This is a theoretically difficult process, and one easily subject to abuse. There is an extensive and complex literature on secondary costs and benefits that specifically addresses this analytical approach.

Instead of adding on secondary costs and benefits, one can either adjust the values used in economic analysis or incorporate the secondary costs and benefits in the analysis, thereby in effect converting them to direct costs and benefits. This is the approach taken in most project analyses carried out by international agencies, in the systems based on shadow prices proposed in more recent literature on project analysis, and in the analytical system presented here. Incorporating secondary costs or benefits in project analysis can be viewed as an analytical device to account for the value added that arises outside the project but is a result of the project Investment. In the analytical system here, every item is valued either at its opportunity cost or at a value determined by a consumer's willingness to pay for the item. The effect is to eliminate all transfers-both the direct transfers and the indirect transfers that arise because prices differ from opportunity costs. By this means we attribute to the project Investment all the value added that arises from it anywhere in the society. Hence, it is not necessary to add on the secondary costs and benefits separately; to do so would constitute double counting.

One qualification must be made. If a project has a substantial effect on the quantity other producers are able to sell in imperfect markets-and most markets are imperfect-there may be gains or losses not accurately accounted for. This diversion entails a social gain from reduced rail traffic (in avoiding the social losses previously incurred on this traffic) in addition to the benefits to the road users measured directly. In agricultural projects, this is a rather infrequent case because prices generally are more flexible than in other sectors of the economy. In any event, in the practice of contemporary project analysis the size of these gains or losses is generally assumed to be insignificant, and no provision is made for them in the analysis.

Although using shadow prices based on opportunity costs or willingness to pay greatly reduces the difficulty of dealing with secondary costs and benefits, there still remain many valuation problems related to goods and services not commonly traded in competitive markets. One way to avoid some of these problems is to treat a group of closely related Investments as a single project. For example, it is common to consider the output of irrigation projects as the increased farm production, since, valuing irrigation water is difficult. Another example is found in development roads built into inaccessible areas. It is argued that the production arising from the induced Investment activities of otherwise unemployed new settlers should be considered a secondary benefit of the road investment. One way of avoiding the problem is to view this case as a land settlement project in which the road is a component. New production is then properly included among the direct benefits of the project and can be included in the project accounts at market or shadow prices, and no attempt need be made to allocate the benefits between road investment and the other kinds of investment that must be made by settlers and government if settlement is to succeed.

Another group of secondary costs and benefits has been called "technological spillover" or "technological externalities." Adverse ecological effects are a common example, and the side effects of irrigation development are often cited as an illustration. A dam may reduce river flow and lead to increased costs for dredging downstream. New tubewell development may have adverse effects on the flow of existing wells. Irrigation development may reduce the catch of fish or may lead to the spread of schistosomiasis. When these technological externalities are significant and can be identified and valued, they should be treated as a direct cost of the project (as might be the case for reduced fish catches), or the cost of avoiding them should be included among the project costs (as would be the case for increased dredging or for investment to avoid pollution).

It is sometimes suggested that project Investments may give rise to secondary benefits through a "multiplier effect." The concept of the multiplier is generally thought of in connection with economies having excess capacity. If excess capacity exists, an initial investment might cause additional increases in income as successive rounds of spending reduce excess capacity. In developing countries, however, it is shortage of capacity that is characteristic. Thus, there is little likelihood of excess capacity giving rise to additional benefits through the multiplier. In any event, most of the multiplier effect is accounted for if we shadow-price at opportunity cost. Since, the opportunity cost of using excess capacity is only the cost of the raw materials and labour involved, only variable costs will enter the project accounts until existing excess capacity is used up. It is also sometimes suggested that there is a "consumption multiplier effect" as project benefits are received by consumers. Consumption multipliers are very difficult to identify and value. In any case, they presumably would be much the same for alternative investments, so omitting them from a project analysis would not affect the relative ranking of projects.

INTANGIBLE COSTS AND BENEFITS

Almost every agricultural project has costs and benefits that are intangible. These may include creation of new job opportunities, better health and reduced infant mortality as a result of more rural clinics, better nutrition, reduced incidence of waterborne disease as a result of improved rural water supplies, national integration, or even national defence. Such intangible benefits are real and reflect true values. They do not, however, lend themselves to valuation. How does one derive a figure for the long-term value of a child's life saved, or for the increased comfort of a population spared preventable, debilitating disease? Benefits of this kind may require a modification of the normal benefit-cost analysis to a least-cost type of analysis, a topic we will take up when we discuss valuation.

Because intangible benefits are a factor in project selection, it is important that they be carefully identified and, where at all possible, quantified, even though valuation is impossible. For example, how many children will enroll in new schools? How many homes will benefit from a better system of water supply? How many infants will be saved because of more rural clinics?

In most cases of intangible benefits arising from an agricultural project, the costs are tangible enough: construction costs for schools, salaries for nurses in a public health system, pipes for rural water supplies, and the like. Intangible costs, however, do exist in projects. Such costs might be incurred if new projects disrupt traditional patterns of family life, if development leads to increased pollution, if the ecological balance is upset, or if scenic values are lost. Again, although valuation is impossible, intangible costs should be carefully identified and if possible quantified. In the end, every project decision will have to take intangible factors into account through a subjective evaluation because intangible costs can be significant and because intangible benefits can make an important contribution to many of the objectives of rural development.

4

Capitalism Economy in Agricultural Production

CAPITALIST ECONOMICS

It is typical of capitalist economics, in contrast to Marxist, to try and put markets and market relations at the centre of its analysis and not production and the class relations immanently inherent in it. The former are looked upon as determinant or self-sufficing, allegedly developing of themselves and exerting a dominant influence on the production process. The idea of the primariness of the sphere of circulation, with very substantial differences in the evaluation of the main moments of the functioning of economic systems of any scale, has dominated capitalist economics throughout its history. True, its original theoretical postulates are now being criticised more and more often by many leading Western economists, people like Samuelson, Galbraith, Myrdal, Erhard, Leontief, Balassa, and Tinbergen.

Still, this idea has continued in fact to permeate capitalist literature in one form or another since World War 11, whether it deals witJi capitalism's internal or world economic problems. In their detailed work on the growth of the international economy, Kenwood and Lougheed declare the central sphere of their study to be that of circulation, since, as they write:

The exchange ol goods and services is the means through which independent economic units enter into economic relations with one another and become part of local or national economic community. An exchange passes beyond a country's boundaries, national economic systems become parts of broader regional, continental or world economy.

Prof. P. T. Ellsworth stresses, in his textbook, which was widely used in the United States and Great Britain in the 50s and 60s, that the very term 'international economies' indicates 'that it concerns itself with the economic relations between nations'. Such a posing of the matter is quite typical of present-day Western economic literature. When investigating capitalism's world economic problems, capitalist economists do not, of course, ignore theoretical analysis of the international aspects of the production process, and study many of them very thoroughly, particularly the dynamics of its productive forces, comparations of the level of economic development of various countries and regions, problems of the international division of labour, capital exports, production and technological competition, integrational processes within the various international economic groupings, migration of labour, and so on.

All these processes, however, arc willy-nilly analysed from ideological positions that pursue definite class interests of the capitalists. The world economy is studied mainly from the standpoint of the determining effect of the trends of world supply and demand on production, and of opportunities for effective employment of the market mechanism of state monopoly methods of regulating international economic relations. Objectively the main job of this approach is 18 to conceal the antagonistic essence of the international relations of production under capitalism. Capitalist economists' endeavour to survey the world capitalist economy out of the context of its organic link with the patterns of development of the corresponding exploiter relations of production, and to represent it simply as a mechanically formed aggregate of national markets within the context of the international exchange of goods and services, also serves the same purpose. On this basis capitalist political economy has time and again exerted great efforts to revive the gimcrack idea whose essence is that modern capitalism is still ahle to generate adequate conditions for future 'harmonious' development of the world economy and in the long run to eradicate its deepest inherent defects and contradictions.

Marxist-Leninist economics considers this matter from fundamentally different theoretical positions. It substantiates in detail the proposition that the world capitalist economy, like a commodity economy of any scale (whether local, regional, or national), includes such inseparably interconnected and jointly subordinated elements as production, markets, and a system of socio-economic relations arising from a corresponding division of labour between economic units and producers. Although market factors of exchange and distribution figure primarily on the surface in capitalism's international economic interconnections, it is not they, however, but the social relations in the production process, that play the basic role when the antagonistic, class nature and decisive trends of development of capitalism's world economic system are investigated.

The capitalist mode of production (Marx wrote) is... a historical means of developing the material forces of production and creating an appropriate world market and is, at the same time, a continual conflict uetwcen this its historical task and its own corresponding relations of social production.

This conclusion has acquired special methodological significance for analysing the deep-seated crisis processes in the modern world capitalist market. It provides an opportunity for a scientific systematising and comprehensive investigation of the objectively operating patterns and irreconcilable contradictions of capitalism's international market relations, which stem from the socio-economic nature of this mode of production.

WORLD CAPITALIST ECONOMY SYSTEM

The decades since the war have been marked by extremely far-reaching changes of world historical significance in all spheres of human affairs. Never before has society experienced such substantial transformations in its socio–economic and political structure in so short a time. Taking them globally, two dialectically linked trends attract attention first of all.

One is the forming and rapid growth of a world system of socialism through the revolutionary replacement of capitalist and pre-capitalist social relations by socialist ones in a number of countries. The other is the steady deepening of capitalism's contradictions within the contracting limits of the non-socialist world. Both these processes provide objective prerequisites for further development of the general crisis of capitalism, an inherent, decisive factor of which is the crisis of the world capitalist economy.

Since World War! II there have been very substantial changes in this economy compared with earlier stages that have affected very important aspects of the economic affairs and international relations of all the countries of the nonsocialist world that comprise it, though by no means to the same extent. Most of them are occurring directly or indirectly under the influence of comparatively new trends in the economic competition of the two world social systems. These trends in turn inevitably reflect the long-term consequences of the break-up of the colonial system, the ever increasing, extremely contradictory influences of the scientific and technical revolution on the productive forces and structure of social production of the various groups of countries, qualitatively new phenomena as regards internationalisation of their social production and the international division of 6 labour, the mounting struggle of the emancipated countries against neocolonial exploitation, for real equality in international economic relations.

The economic and social instability of the world capitalist system became unusually intense at the end of the period we study, especially as a result of the marked sharpening in the 70s of the energy, raw material, inflation, cyclic, international financial, and other major problems of world capitalist economy. These crisis processes, which had a logical continuation and development in the years following, developed in a complicated, eventful situation of intense struggle between two lines in world politics, *viz.*, on the one hand, that of curbing the arms race, consolidating peace and detente, and defending the sovereign rights and freedoms of nations pursued by world socialism and the whole peaceloving mankind, and, on the other hand, the imperialist line of undermining detente and stepping up the arms race, and policy of threats and interference in others' affairs and suppression of the national liberation struggle.

Capitalism, as was stressed at the 26th Congress of the CPSU, has not, of course, congealed in the present conditions of its general crisis; from the beginning of the 70s through to the early 80s alone, it has experienced three economic recessions in spite of capitalist governments' counter-crisis measures and all their efforts to prevent new slumps in production. Inflation has attained an unprecedented scale in these countries. As the Central Committee's report to the Congress noted:

It is more than obvious that slate regulation of the capitalist economy is ineffective. The measures that bourgeois governments take against inflation foster stagnation of production and growth of unemployment; what they do to contain the critical drop in production lends still greater momentum to inflation.

The more that facts are accumulated revealing critical trends in, and specific features of, the growth of the productive forces in the economies of the main groups of countries of world capitalism, the more pressing is it becoming to systematise, generalise, and interpret them in the aggregate and in their inter-relationship. A problem of fundamental importance is arising, *viz.*, whether a complex approach to the tackling of such a broad task is possible at all. For the aggregate of facts on the economic activity of all countries and of all the industries of world capitalism (in their dynamics and interconditioning, and their repeatedly intersecting national and international connections) embraces a truly immense range of matters, each of which plays a far from identical role in the development both of the national states, and the industries and spheres of their economies, taken separately, and of the world economy as a whole.

The answer to this question was first given by Marxism; and it consisted not so much in the discovery of some allembracing method for summarising a vast number of facts and trends as in its bringing out the theoretical and methodological premises of a systems approach to analysis of the inner logic and patterns of the capitalist system's functioning at the various concrete stages of its 'self-movement'. Lenin stressed, when speaking of the changes constantly occurring in the capitalist world economy, that the sumtotal of these changes in all their ramifications in the capitalist world economy could not bo grasped even by seventy Marxes. The most important thing is that the *laws* of these changes have been discovered, that the *objective* logic of these changes and of their historical development has in its chief and basic features been disclosed. It is its understanding of the objective laws of social development that has enabled Marxist-Leninist economic science to tackle elaboration of the principles of a scientific methodology for analysing capitalism as a world economic system.

CAPITALIST MODE OF PRODUCTION

The characteristics of the most important stages of the long forming and evolution of this system taken together with the genesis of the capitalist mode of production, the world market, and the aims and methods of colonial exploitation of some countries by others, constitute the main substance of Part I of this book, in which special attention is paid to defining the object of study and the internal contradictions and long-term trends of its development that shape the basic causes of the rise and inevitable deepening of the crisis of the world capitalist economy today. The data adduced will provide the necessary theoretical foundation for bringing ont the themes of the succeeding chapters.

The general results and decisive trends of the development of the postwar world capitalist economy, all the ecanomic factors and social consequences of the long-term shifts in the structure and dynamics of both the social production and the consumption of all the countries of this economy taken together.

Then follows a detailed comparison of two groups of countries (*viz.*, capitalist and developing) in order to bring out the specific character of their postwar development, including the impact of the scientific and technical revolution and break-up of the colonial system on their economies. The trends in the growth of population and manpower have been analysed from the same standpoint, and also the productivity of social labour in the main geographical regions of the non-socialist world. All this has enabled us to attempt a more circumstantial and detailed estimate of the changes that have come to light in the regional structure of the distribution of the leading sectors of social production in both the industrial centres and the agrarian and primary producer periphery of capitalism. The problem of the cyclic movement of the postwar world capitalist economy has a place of fundamental importance in the book. Analysis of it helps clarify many features of the regular intensification in recent decades of the uneven economic growth and structural instability of modern capitalism's international business relations.

The main lines of the intertwining development of the industrial and agrarian and raw material base of the postwar world capitalist economy are analysed in Part III. Study of the changes in the manufacturing industries in capitalist and developing countries that are most significant as regards their socio-economic consequences occupy the foreground. Then the long-term shifts in the dynamics, volume and structure of the production of raw materials and in the movement of prices of industrial items, farm produce and raw materials on the world capitalist market are examined. On that basis a number of objectively operating trends are brought out, in which the essential features of the present stage of the crisis of the imperialist system of international division of labour are being more and more clearly manifested within the framework of the world economic relations of recent years. It would not be legitimate to appraise all the trends studied in isolation from the decisive patterns of the world 9 revolutionary process of modern times. As Leonid Brezhnev stressed at the 25th Congress of the Communist Party of the Soviet Union: This is an epoch of radical social change. Socialism's positions arc expanding and growing stronger. The victories of the national liberation movement are opening up now horizons for countries that have won independence. The class struggle of the working people against monopoly oppression, against the exploiting order, is gaining in intensity. The scale of the revolutionary-democratic, anti-imperialist movement is steadily growing. Taken as a whole, this signifies development of the world revolutionary process.

This book does not claim, of course, to be an exhaustive analysis of the problems posed; many of them certainly need further study, tidying up, and concretisation. Because of the inadequacy or scrappiness of reliable statistics and general economic data, separate propositions of the themes are reviewed in outline or are advanced tentatively. When the conclusions are based on published work or well-known data the exposition has been compressed to a resume, and references are given to the sources containing the detailed argumentation.

The Russian original appeared in 1978. In the present English edition the author has tried to continue analysis of the problems and trends studied, employing the latest data available to him. On that basis certain amendments have been made, and the conclusions and estimates of the probable outlook for the development of the world capitalist economy have been elaborated and broadened.

PRECEDING COLONIAL DEVELOPMENT OF CAPITALISM

During the whole preceding colonial development of capitalism, a decisive part of the aggregate product of its periphery was, as we know, created in agriculture, which was a natural consequence of the underdevelopment of the social production of the colonies and semi-colonies. This underdevelopment was further deepened by the fact that the colonial authorities saddled them with the role of economic appendages of the metropolitan countries. Their agrarian production was forcibly oriented on supplying the centres of capitalism with produce at the expense of their own consumption, so that while agriculture was the main occupation of the bulk of their populations, hunger and poverty prevailed in them, which was one of the most pernicious consequences of the transformation of capitalism into a world economic system.

By the middle of the twentieth century the gap ih this sphere between the metropolitan countries and their agrarian, primary commodity producing appendages had attained colossal proportions. The weight of agriculture in the latter's GDP was then almost double that of industry and constituted nearly 90 per cent of the value of the production of all sectors of the circulation and services sphere put together. In the metropolitan countries, however, the weight of agriculture was correspondingly roughly a quarter of that of industry and 87 per cent less than the proportion of the services and circulation sphere. Since then there have been notable changes in the structural proportions, which is very important to allow for not only when estimating the shifts that have already occurred but also when calculating probable coming shifts in the distrubition of the world economy's productive forces. As for the developing countries, it is worth noting that agriculture, whose share still considerably exceeded that of the industrial product at mid-century, has since finally and irrevocably fallen to second place behind industry. There are real grounds for predicting a further change in this respect. In our estimation the line of development noted may lead to the weight of industry in the GDP of developing countries being at least double that of agriculture by the end of the 80s. Their lag behind the industrialised countries, however, will still undoubtedly be very considerable then, since the industry/agriculture ratio in the latter was already estimated at around 8.5 : 1 in the 70s, and will probably not be less than 9 (10) : 1 in the 80s.

Comparison of the long-term trends of development of industries and direct services also leads to not uninteresting estimates of the changes occurring, or anticipated in the future, in the structure of the GDP of the two groups of countries. According to them the ratio between agrarian production and the services sphere, for instance, steadily changed in favour of services in the economic centres of capitalism. At the beginning of the 80s 14 times more commodities were produced in it (in terms of value) than in agriculture. A similar process, but at a qualitatively different level, was observable in the last decades in the former colonial world as well. Even at the beginning of the postwar period the share of services in the aggregate product was around 60 per cent less than that of agriculture. Three decades later it had already risen much above the latter. To all intents and purposes, however, the services sphere will retain a clearly marked tendency everywhere in the foreseeable future towards priority development compared with agriculture, probably with accelerated growth rates where the backwardness of the relations of production previously maintained by colonialism has long been an obstacle to its establishment and development.

Comparison of the growth results of the other main spheres of material production (*i.e.* industry and the sectors constituting 'services') yields a diametrically opposite picture for the two groups. At the beginning of the postwar period the proportion of industrial production in the GDP of both of them (especially of the second group) was less than that of sectors belonging to the services sphere. But in recent decades the proportion of industry came noticeably closer to the aggregate share of the services sectors and later overtook them in both groups of countries. In other words, there was a smoothing out in this respect, too, of the acute imbalances in the structure of social production in both groups with the decay of the colonial system, while the one preserved its immense superiority over the other as regards level of economic development. We shall shortly demonstrate below how the structural proportions between the principal industries within each group gradually altered after the war, but already, on the basis of the facts cited, we can conclude that, in spite of the extremely broad variety of the trends of socio-economic development, the determinant features of the postwar dynamics of the growth of social production in the countries taken separately, and the structure of the distribution of production created by capitalism between its industrial centres and agrarian periphery, are beginning to undergo fairly important changes.

As the graphs indicate, there has been a long-term trend in literally all the principal sectors forming the gross product of the non-socialist world towards an enhancing of the role of the production of developing countries. In the period covered by their share in the industrial output of the capitalist economy has increased by roughly 75 per cent, in building and construction has doubled, in transport and communications has risen by a third, in services by twothirds, and in trade by more than a quarter. That does not, however, mean any kind of fundamental break in the longstanding extreme unevenness of the distribution of production capacities between industrially developed countries and the primary producers in the present day capitalist economy. In all its principal sectors, except agriculture, five to seven times as much output originated in the metropolitan countries at the end of the 70s as in the developing ones. Even the indisputable fact that the developing countries surpassed the main economic centres in volume of farm production does not give us the right to consider there to have been any kind of essential turn in the distribution of the productive forces. Furthermore, it is agriculture that is still, as a rule, the most backward sector of their economies, which to a crucial extent determined the steady widening of the gulf between them and the developed countries in per capita gross production.

THE AGRARIAN AND RAW MATERIAL PRODUCING COUNTRIES

The winning of political independence by the bulk of the countries of the colonial world in the postwar period opened up fundamentally new opportunities for them to fight to step up their national development, and for socio-economic progress and equality in international relations against imperialist diktat and exploitation by expatriate monopoly capital. As Leonid Brezhnev said at the 25th Congress of the CPSU: Glancing at the picture of the modern world one cannot help noticing the important fact that the influence of states that had only recently been colonies or semi-colonies has grown consider ably.

It may definitely be said about the majority of them that they are defending their political and economic i ights in a struggle against imperialism with mounting energy, striving to consolidate their independence and to raise the social, economic and cultural level of their peoples. In these conditions important changes are occurring or coming into life in the structure and distribution of the developing world's social production, which are putting their stamp on the processes that govern the peculiarities of the former colonies' and semi-colonies' position in the present international division of labour, including their mounting struggle to lay the foundations of their economies. In the 50s through the 70s the physical volume of their industrial production as a whole rose by more than 450 per cent, so that now, in contrast, to the earlier stages of their economic history, industrial goods, especially manufactures, are beginning to predominate in the material production of most of them. At the end of our period somewhat more was produced by manufacturing industry in value terms (in 1970 prices) than in agriculture, and roughly four times as much as in mining.

As the general economic capability, and especially the industrial potential, of these countries grew there was an inevitable consolidation of the demarcation of class forces, a deepening of the heterogeneity of their social development, and a growth of class struggle. For the immense majority, at the same time, that were to one degree or another within the orbit of the laws of the world capitalist cconomy, it remains characteristic that the tasks of the anti-feudal revolution have not been completed and the socio-economic survivals of the colonial period are very burdensome. Many of the new sovereign states had begun to carry through radical reforms during the liberation struggle that were both antifeudal and anti-capitalist. In thatsituation the differentiation of their paths and levels of economic development greatly broadened.

In spite of essential differences of that kind, however, all the emancipated countries are part of the agrarian and raw material periphery of the industrial centres of capitalism. Even the most economically advanced of them are still mainly exporters of raw materials arid importers of machinery, equipment, and other industrial goods. They 158 continue to he in an unequal position within the capitalist world system and subjected to exploitation by expatriate monopoly capital, which is why they are being urgently faced with a number of common internal and external economic problems.

PRODUCTION DISTRIBUTED AMONG THE DEVELOPING COUNTRIES

The most important trends in the changing postwar structure of the periphery's social production have been analysed above. How, however, was this production distributed among the developing countries in the first postwar decades? As will be clear from the graphs the growth rates of aggregate production were very fast in Latin America, where most of the countries had achieved national independence and taken the road of independent capitalist development long before break-up of the colonial system began. As a result the postwar years have been marked by a gradual raising of the role of Latin America in the aggregate product of the developing world, from 40 to 44.5 per cent, so that it is now greater than that of the Asian region although the latter had nearly four times its population at the end of the period surveyed.

This growth did not come about through the contribution of all Latin American countries, but occurred on a background of a marked imbalance of their economic development and was mainly determined by a few leading countries, primarily Brazil and Mexico, for which much higher growth rates were typical. At the end of the 70s more than 60 per cent of the aggregate product of the region was produced in these two countries alone, whereas they produced around 40 per cent of it at the beginning of the 50s. At the same time certain countries (Haiti, Honduras, Barbados, etc.), still caught up in heavy chains of dependence on expatriate commodity monopolies, remained among the most economically 159 backward appendages of the world capitalist economy, with the usual low growth rates typical of semi-colonial countries. A steady deepening of the unevenness of Hie intor-regional. and especially of the intraregional, distribution of production has become typical of the various groups of commodityproducing countries in the other continents. In Africa, where colonialism succeeded in retaining key positions in many areas for a number of years in the postwar period, the course of economic development has proved to be relatively slow, so that the weight of the continent in the aggregate social product of developing countries has had a clear tendency to fall, despite the fact that the total GDP of African countries has risen substantially.

As national independence has been consolidated and African countries have thrown off the yoke of colonial exploitation, however, their growth rates, it must be noted, as in other areas of the developing world, have gradually risen. They have been highest, as a rule, in the Arab countries in North Africa, where the positive consequences of the breakup of the colonial system madc themselves felt much earlier than in Tropical Africa. The rapid increase in income from exploiting their oil wealth, moreover, provided some of them with favourable material conditions for their struggle for quicker economic development.

There has been an ever wider difference in levels and rates of development in Asian countries, where around two-thirds of the developing'world's population lives. Most of Ihem, which were in forced dependence on a handful of European powers during nearly the whole history of the forming of the world capitalist economy, were able to win national independence at the start of the postwar period. In the 50s they had already begun an active fight for accelerated growth of their previously stagnating productive forces. The mean annual growth rates of the GDP of all the developing countries of Asia reached 4.5 per cent in that decade, 5.1 per cent in the 60s, and around 6 per cent in the 70s. This accelerated growth was due to no small degree in recent times, however, to the 'oil boom' in the Near and Middle East. The real volume of the GDP (in value terms) increased more than sixfold in that region in the 50s through the 70s, and was more than one-third of the GDP of all the developing countries of Asia, although its population at the end of the period was only 7 per cent of the continent's total. At the same time the GDP of all the other developing countries in South and South-East Asia, which have immense human resources, rose less than threefold. If we examine these processes through the prism of the data on population growth, the generalisations made above, stemming from a comparative description of the regional dynamics of production, obviously need to be made more precise, and corrected. The successive changes in the indicators of per capita GDP for the separate regions and countries enable the scale of the shifts in the geographical distribution of their productive forces to be brought out more clearly.

In almost all the groups of countries in the table, there is a quite marked tendency, albeit extremely varied in dynamics and significance, towards a differentiation of population growth rates, in addition to a sharpening of the differences in dynamics of production. This trend characterised one of the inevitable consequences of the break-up of the colonial system, and could not help having an essential influence on the ratio of economic potentials, and consequently on the mutually interlocked processes of economic development. In the Asian region a particularly wide gap has been formed in the dynamics of per capita production of the aggregate social product between the oil-producing countries of the Near and Middle East and the overwhelming majority of the countries of South and South-East Asia.

Similar processes have been occurring in Africa, for all its special features. The growing uncvenness of development has been deepened there, even more than in Asia, by considerable differences in rates of population growth. In some cases these differences, it is true, have encouraged a convergence of the dynamics of per capita production in the separate groups of countries. That is confirmed, in particular, by comparison of the lines of postwar development of North and Tropical Africa. Population has grown faster in the former, which has led to a certain levelling out of the aggregate indicators of per capita growth of production used in the table. At the same time there have been quite different trends within each of these regions.

In the biggest North African country, Egypt, for example, there was an obvious slowing of the growth rates both of production and of population in the 60s and 70s. The annual rates of per capita growth of GDP in those years were 1.5 per cent in Egypt and only 0.2 per cent in Sudan. In other words, they were lower than in many of the countries of Tropical Africa whose economies on the whole still suffer maximally from the burden of harmful survivals of the colonial period compared with other developing countries. Although some of the new states south of the Saharajiave achieved a relatively high growth of production (Kenya, Gabon, Ivory Coast, Nigeria, Malawi, etc.), most of them still have extremely slow and very unstable rates of development. In quite a few countries, moreover, the main indices of the dynamics of GDP growth have been lower in recent decades than the growth of population (Burundi, Chad, the Central African Empire, Uganda, Upper Volta).

TRENDS IN THE DISTRIBUTION OF PRODUCTION

The long-term trends in the distribution of production in the former colonial regions are primarily governed by the resultant tendencies of the development of a few of the biggest countries (as regards area, population, and natural 164 resources). The ratio of their economic capabilities is therefore important in a generalised description of intraregional processes, including the contradictory consequences of the 'demographic explosion' in the Third World. In Latin America, for example, GO per cent of the inhabitants and nearly two-thirds of the aggregate GDP of the whole region were concentrated in Brazil, Mexico, and Argentina. Among the 'Big Three' production developed at the lowest rates in Argentina in the postwar years, but the comparatively slow dynamics of its population growth led to its per capita production, for example, exceeding Brazil's in the 60s and Mexico's in the early 70s. It is typical that the economies of several small countries (Uruguay, El Salvador, Paraguay, Haiti, Honduras, and others) wore below the average of the developing countries of South and SouthEast Asia as regards the same indicators in those years, not to mention of bulk of the emancipated countries of Africa. The examples given, whose number could be greatly extended, very clearly confirm that there has been a steady sharpening of the unevenness of economic development of the former colonies and semi-colonies in recent decades.

This irresistibly intensifying process, however, in no way refutes the need (in addition to the study of the features of the economic growth of each country) for a comprehensive systems approach to analysis of the summary trends of their interconnected development within the whole developing world.

Such an approach helps, on the one hand, to bring out the objectively operating trends in which the extremely contradictory and isolated processes and phenomena characterising the shifts in the distribution of the productive forces of the separate countries and their regional groups are ultimately reduced, and on the other hand, gives the possibility of better understanding and appreciating the specific features of their development.

In this connection comparative analysis of the long-term changes in the industrial structure of the social production of countries in Asia, Africa, and Latin America is becoming of paramount importance. As these countries are industrialised this process will undoubtedly hec07iie more and more pronounced.

Its effect gives grounds for supposing that even before the end of this century the share of agriculture in the GDP of most of the developing countries of Asia and Africa may fall by 50 to 50 per cent and reach the level attained in Latin America al the beginning of the 80s. At the same time the proportion of agriculture in the GDP of the Latin American region as a whole will also probably diminish, gradually approaching the present average level of the industrial capitalist countries (around 4 per cent). An inevitable consequence of that will be the pushing of increasing numbers of the rural population into the cities.

Another very important trend determining botli the current structure of changes in the production of the developing countries and their probable future structure has become the general though not uniform rise-in the role of industry in their GPP. The scale of its growth, moreover, has been quite regularly higher in the economically more backward countries. Thus the proportion of industrial output in the aggregate product for the period considered in the graphs increased in the African region by 120 per cent (from 12 to 26 per cent), in the Asian region by almost 100 per cent (from 14 to 27 per cent), and in the Latin American region by almost a third (from 24 to 31 per cent).

As a result of the increasing imbalance of this growth, a trend towards a certain convergence of the significance and place of industry in the general structure of these regions' production has been noted in the postwar period. The proportion of industry in the GDP of countries in South and South-Easl Asia and in Africa was 50 per cent less than in Latin America at the beginning of the 50s; at the beginning of the 80s the gap had noticeably closed and was estimated now at 1:1.3 for the former and 1:1.2 for tho latter.

This trend will intensify, in all probability, in the future, and characterise one of the tendencies of developing countries' uneven economic growth. It can be expected that the weight of industry in the GDP of most Asian and African] countries will come close to tho present level in the leading Latin American countries in the next two or three decades, while the latter will reach the average level of the present-day industrial centres of capitalism.

These very general estimates provide the dnta needed, but only the initial data, for studying the significance and place of industry in the regional economies. They naturally smooth over tho vast differences in structure and sectoral dynamics of the industrial production of the separate regions and countries.

A more and more concrete analysis of the separate industries is a central task of the part that follows, but here we must bear in mind that these processes of industrial development in agrarian and raw material producing countries are largely linked with rapid growth of mining, whose output is mainly sold on the world capitalist market, and in fact is still really controlled to one degree or another by expatriate monopoly capital. The rapid growth of the mining of minerals, and in particular of fuel, has played no small role in the regions being considered in the postwar rise in the proportion of industry in the gross product.

It would not be legitimate, however, to conclude from this that the steady raising of the role of industry in the gross product of developing countries is more or less wholly 168 linked with the working of minerals. The rapid development of their manufacturing industries in recent decades has had an ever increasing effect, especially the development of heavy industry, the volume of whose production rose on the whole by almost 150 per cent in 1968–80.

At the beginning of the 80s the heavy industries of all developing countries were producing 30 per cent more output in value terms (in added value in 1975 prices) than the light industries. There is little doubt that their basic industries will also grow al faster rates in the next few decades, which will further a gradual raising of their role in industrial production.

For a comparative estimate of the long-term trends bringing out both the general and the particular in the production sphere of the main developing regions of capitalism, it is essential to compare the mean annual growth rates of their main industries (including per capita rates) over some considerable period. Because of the inadequacies of the economic statistics of developing countries it is very difficult to bring out these growth rates in summary form (in contrast to developed countries). The comparison can only be attempted on the basis of very approximate calculations, which allow us to characterise the most general trends of the sectoral shifts with a certain degree of accuracy. At the same time they can be used as a starling point of sorts for studying the specific features of the development of production in the separate countries and groupings in each of the regions.

In agriculture, which is still in a slate of deep, protracted depression in the developing world as a whole, Latin America and most of the countries of the Near and Middle East have displayed relatively faster growth, so that a line towards an increase in their share in developing countries' total production of farm produce has begun to show., with a corresponding drop in the role of South and South-East Asia and Africa.

As a result they have become further differentiated as regards per capita farm production. By the beginning of the 80s the Latin American countries on the whole considerably surpassed the countries of Asia and Africa in this respect. It is unlikely that there will be any closing of this gap in the near future.

The fight to eliminate the extreme backwardness of agriculture will undoubtedly remain one of the most pressing socio-economic problems of most of the countries of the former colonial world in the coming decades. It is the slow development of their agriculture, compared with population growth, that is the decisive cause of the consistent tendency manifested since the war towards a widening of the gap in GDP per capita between the developing and developed countries of the capitalist world.

In the 60s and 70s, however, this trend began to be more and more actively countered by processes developing in most of the other sectors of the economies of the former colonial periphery of capitalism, and primarily in industry. The international statistics clearly indicate that the per capita growth rates of industrial production in economically most backward regions of capitalism's periphery have noticeably exceeded the industrial centres in per capita growth rales of industrial production. In the developing countries of Asia per capita industrial production increased by a factor of 4.3 from 1950 to 1978, in Africa by 3.5, and in the developed capitalist countries taken together by 2.6. The countries of Latin America also showed a rather faster growth in this respect, on an average, than the main economic regions of capitalism.

In present conditions there has begun lo be a tendency towards a certain closing of the gigantic, gap in per capita industrial production formed under colonialism, together with 171 a gradual increase in the weight of the agrarian and raw material producing countries in the industrial production of the non-socialist world. In Hie very early 50s this gap was estimated at 1:35 for Asian countries, 1:32 for Africa, and 1:5.5 for Latin America. By the end of the 70s it had been clearly reduced and was expressed by the following approximate ratios: Asia 1:24, Africa 1:27, and Latin America 1:5.3. In the middle of 70s the substantial slowing of industrial growth rates in the developed countries connected with the greatest postwar cyclic crisis of the capitalist world economy fostered a consolidation of this trend.

The general course of this process leads to th,e conclusion that it is acquiring a quite stable, irreversible character in Ihc present situation. Being a natural consequence of the break-up of the colonial system, it has already begun to have (and in all probability will continue to have) a growing impact on the steady deepening of the crisis of the unequal system of international capitalist division of labour built up on a colonial basis.

At the same time the facts adduced confirm the extreme complexity and difficulty of the problem of overcoming the age-old industrial backwardness of the former colonies and semi-colonies, especially with their constant exploitation by expalriate monopoly capital. The solution of such a major and objectively inevitable hislorical task will obviously require several decades and will continue for a long time to be one of the urgent problems ol the light of the peoples of the liberated countries for final liquidation of the heavy legacy of the colonial period and imperialist domination of their economies.

To reduce their backwardness by even two-thirds from the present mean per capita industrial production of the developed countries, for instance, it would take the countries of South and South-East Asia around a quarter of a century at the rates of real industrial growth of the 60s and 70s, Africa more than a quarter of a century, and Latin America around 30 years. Then, however, per capita output of industrial goods in each of these regions would the respectively one-eighth, a ninth, and about a third below the level already reached by Ihe advanced capitalist countries at the end of the 70s. If that level remains unchanged in the future, then in that hypothetical case the way out for most developing countries would have lo be a 15-fold or 16-fold expansion of the volume of industrial production within the present-day world capitalist economy. These figures cannot, of course, serve as a generalised criterion of a quantitative, let alone a qualitative, estimate of the future parameters and goals of the industrial development of all the developing countries of Asia, Africa, and Latin America. In fact, however, they help determine only the scale of the industrial backwardness of developing countries. It inevitably follows from them that the struggle for a cardinal solution of the problem in the present transitional epoch will give rise to an urgent need to develop a long-term strategy and-strictly scientific approach to the global outlook for industrialisation of the former colonies and semi–colonies on quite other principles of social and world economic relations compared with those historically built up by capitalism.

The postwar processes in the agrarian and industrial spheres have consequently had a different effect on the resultant ratio of per capita GDP in the groups of countries under consideration. In the agrarian sphere they have fostered a widening of the economic gap, and in the industrial sphere a narrowing. But since the role of the agrarian sector in the GDP of developing countries is more significant than in the centres of capitalism, and the industrial sector less significant, the gap between them in per capita output of the main sectors (agriculture, industry, and building) has also continued to widen in recent decades, and for Asian and African countries was more than 10:1 at the beginning of the 80s against 8:1 in the first postwar years. In other words, at the turn of the decade to the 80s, per capita production in value terms (in comparable prices) was much lower in developing countries compared with capitalist ones, than at the turn of the 40s and 50s.

INDUSTRY GROWTH IN DEVELOPING COUNTRIES

As for industry, its growth in developing countries has still not led to any real qualitative shifts in the liquidation of age-old industrial backwardness. The struggle to lay an industrial foundation for their national economies is still more or less, as a rule, in the initial stage. In spite of a narrowing of the gap in per capita industrial output, they now produce dozens of times less product than the average for capitalist countries.

The most powerful, technically advanced branches of production, moreover, that can more efficiently exploit the fruits of the present scientific and technical revolution are concentrated in the latter. That is affecting the dynamics of productivity in the industry of both groups of countries. In the 60s and 70s its annual growth rates in the economic centres of capitalism were double that in the developing countries.

There has been a quite stable tendency towards a gradual, tliough slow, rise in the proportion of circulation and services in the GDP of the developing world, which rose perceptibly in 1900–80. As a result of the rapid expansion of this sphere its weight in the GDP of the Asian and African regions rose by a quarter and nearly a third respectively in the 50s through the 70s. It rose a little as well in Latin America on the whole and constituted more than half of that region's GDP at the end of the 70s. In other words, there was a certain convergence of structural proportions in the sectors of non-materialproduction of the main developing regions, but the historical underdevelopment of these sectors still remains a distinguishing sign of the economic backwardness of the vast majority of the former colonies and dependent countries of these regions.

The proportion of developing countries in the aggregate production of the services and circulation sphere of the world capitalist economy is beginning lo grow. At the start of llio 80s it was 10 per cent against 11 per cent in the early 50s. The per capita gap between them and the developed countries, it is true, not only did not in fact diminish but even widened a bit on the whole, and was estimated at 15.5:1 in the late 70s, *i.e.* was considerably wider than in the sectors of material production.

For the main sectors of the circulation and services sphere this gap has not built up in the same way; in practice it has always been narrower in trade. But it is in that sector, as in agriculture, that the capitalist countries continued steadily to outpace the developing countries in per capita indices of growth, so that, although the proportion of the latter in the total value of the home trade of the capitalist world lias risen since the war (from one-seventh to one-sixth), there has been another trend in per capita terms While the value of per capita trade (in constant 1975 prices) in the GDP of capitalist countries was 12 times as much as in developing countries at the beginning of the 60s, it was 16 times as much at the start of the 70s.

This trend above all determined the effect in the circulation and services sphere of the tendency towards a further widening of the general economic gap in per capita GDP between the two groups of countries. The intensification of this trend also to no small extent was promoted by processes taking place in the sectors producing so-called services.

Developing countries' backwardness in this respect has long been deeper than in the other sectors of the sphere of social production we are considering. At the turn of the 50s and 60s it corresponded approximately to the level of their industrial backwardness. At that time less than one-tenth of the value of the services created in all capitalist countries was produced in developing countries, and the gap per capita was nearly 20:1.

In recent years there have been changes in various directions in these ratios. The rapidly increasing dynamics of the growth of services in developing countries has led to their share rising in absolute terms to one-seventh at the end of the 70s. But in per capita terms the gap continued to widen for a long time, and in the early 70s could be expressed as 19.5:1, which was mainly due to the backwardness in this 175 area of the economically less developed and more densely populated countries of the African and Asian regions. In the second half of the 70s alone, especially during the world cyclic crisis of 1974–75, a long-term consequence of which was a marked slowing of the growth rales of services in the centres of capitalism, did this gap begin to narrow a bit; at the beginning of the 80s it was expressed by a ratio of 17.5:1 (in 1975 prices). In other words, it has remained wider than in industry. The extreme backwardness of the services sphere governs several of the paramount aims of the long-term socio-economic strategy being developed by developing countries today in their striving to attain the maximum possible speeding up of the rates of social progress in present-day conditions and to raise the standard of living of their populations.

The only major group of services and circulation sectors in which developing countries managed to get a relatively stable increase in the dynamics of per capita production over that of the developed countries in the postwar decades was transport and communications; the grand total of per capita production in them increased by a factor of 3.4 against 2.4 in the developed countries.

There was a corresponding ri se (from 11 to 16 per cent) in developing countries' weight in the aggregate value of the product of transport and communications in the world capitalist economy. In per capita terms their mean annual rates of development have been higher of late than those of the industrial regions, so that in the area of the infrastructure, and in industry, there has begun to be a relatively new and apparently promising trend towards a certain narrowing of the gap in per capita indices, though it still remains extremely wide. If the present dynamics of the growth of production and population is maintained the gap should be halved at least on an average in the coming two or three decades in each of the main regions of the developing world, but its ultimate closing will undoubtedly still remain an urgent task for a long time.

At the same time there was a clearly increasing unevenness in the development of transport and communications in the developing countries, which, as the figures given above show, 176 was organically characteristic of all the other sectors of tho circulation and.services sphere. Its dynamics will also undoubtedly be a close function of the degree and scale of development of material product ion and of the population grovyth rates in the various regions and countries of the developing world.

On the whole, however, one can say with confidence that as the productive forces rise the weight of this sphere in the social production of these countries will gradually increase although it will still lag greatly behind the corresponding average indicator for the industrialised countries for several decades.

The wide gap between the developing countries and the centres of capitalism in all the decisive areas of economic activity consequently remains one of the objective realities of our time. Formed long ago, during the building of the world capitalist economy on a colonial basis, it now characterises many of the most important features of that economy, and of tho whole system of its postwar international relations.

The developing countries of Asia, Africa, and Latin America, given the existing differences in their levels of sociopolitical and economic development, arc inevitably coming up against a need to deal with several common problems, vitally important for all developing countries, on both an international and a national scale. They include the very intricate set of each country's national problems, but ultimately tho struggle to cope witli all of them, taken together is objectively directed to eliminating the pernicious surviv' als of the colonial period from the affairs of human society" and to emancipating the developing countries from neoco, lonialism and continuing capitalist exploitation.

At the same time, as will be clear from the facts adduced above, the gap in levels of economic development remains extremely wide, in spite of considerable progress by the developing countries in coping with these problems, and moreover retains a tendency to widen in several determinant indicators, above all in gross product per capita. The relations of inequality and rapacious exploitation of the natural and human resources of the developing countries by expatriate monopoly capital prevailing in the world capitalist economy are furthering maintenance of this trend, even since the break-up of imperialism's colonial system.

The capitalist social system itself not only bears responsibility for the socio-economic backwardness of the peoples of the former 177 colonial world, but the exploiter system of international division of labour created by it largely continues today to fetter their productive forces. The basic patterns of the world capitalist economy still operate in the direction detrimental to the developing countries.

The striving of the developing countries to spread the liquidating of colonialism to the economic sphere, to put an end to their exploitation by the industrial powers of the West, and to achieve creation of the necessary external and internal conditions for attaining a level of development in the foreseeable future corresponding to the needs of today, is quite justified. At the same time all the nations of the earth without exception have an interest in the speediest attainment of these aims. The struggle for a steady acceleration of socio-economic and cultural progress in the former colonies and semi-colonies is therefore becoming an inseparable part of the fight of all progressive mankind for detente and the consolidating of lasting peace, for further development of the scientific and technical revolution for peaceful purposes, and for a favourable solution of the other global problems of today on which the whole subsequent course of world development to an enormous extent depends.

In his day Karl Marx concluded that mankind thus inevitably sets itself only such tasks as it is able to solve, since closer examination will always show that the problem itself arises only when the material conditions for its solution are already present or at least in the course of formation.

In the present historical situation more favourable objective conditions are being created than ever before for the countries of Asia, Africa, and Latin America that have liberated themselves from the imperialist yoke, or are in the course of doing so, for a successful struggle to overcome the backwardness inherited by them from the past in all sectors of 178 the economy and oilier areas of social affairs.

But every kind of obstacle is still being put in llio way of this bard, long, stubborn light with the world capitalist economy by the former colonialists and expatriate monopoly capital. In these conditions the developing countries are uniting their efforts more and more firmly in the anti-imperialist movement to establish a new international economic order.

The Soviet Union and other socialist countries are supporting these efforts in every way possible and demonstrating their solidarity with the progressive aims of the struggle of the peoples of the developing world for a better future. As the Central Committee's Report to the 26'th Congress of the CPSU stressed: the CPSU will consistently continue the policy of promoting cooperation between the USSR and the ncwly-freo countries, and consolidating the alliance of world socialism and the national liberation movement.

The long-term trends reviewed in this chapter thus not only help us better to understand some of the general results of the changes that have already taken place in the structure of the distribution of the productive forces of the two groups of countries of the postwar capitalist world, but also to distinguish sqveral objective premises for analysing the probable outlook for subsequent structural shifts leading to aggravation of the internal contradictions and crisis phenomena in the capitalist economy. Study of these trends obviously calls for a need to allow in every way for the effect on them of the spontaneously operating laws of the cyclic development of the capitalist mode of production.

POSTWAR CAPITALIST WORLD ECONOMY

The growth of human resources in the much contracted postwar capitalist world economy has been unparalleled in scale. The population of all its countries increased by more than 1,350 million in 1950–80, which was more than double the corresponding figure for the first half of the century (620 million). As a result the total population of the capitalist and developing countries was nearly 3,000 million at the beginning of the 80s (as against only a little more than 1,600 million at the beginning of the 50s and 1,000 million at the beginning of the century). In other words, human 119 resources increased by 80 per cent between 1950 and 1980 within the non-socialist world, and by roughly 190 per cent since the beginning of the century. These considerable changes were determined to a crucial extent by the 'demographic explosion', whose causes are linked with the break-up of the colonial system and advances in health protection. These points have been widely discussed in Soviet socio-economic literature, so we shall touch here only on those consequences of population growth that characterise its distribution in the principal regions of the capitalist world and are essential for understanding the shifts in the distribution of manpower among them, and consequently of the long-term trends in the development of their basic productive force.

Such an immense disproportion has not, by any means, been typical of the whole period of the development of capitalism's world system. Estimates based on UN demographic statistics indicate that even at the beginning of this century the growth rate of the metropolitan countries was patently higher than in the colonies and dependent countries of Asia, Africa, and Latin America, where 120 epidemics raged and hunger and high infant mortality prevailed. The expectation of life in the latter was about half that in the metropolitan countries. Since the war there has been a marked exacerbation of the demographic situation in the capitalist %orld. The rates of increase began to rise in the former colonial world. In 1938–50 they averaged 1.4 per cent a year, in the 50s 2.3 per cent, and in the 60s and 70s 2.6 per cent. In the developed capitalist countries the rates were also rather higher in the first postwar decades, but again began to fall and in the 70s were around 0.85 per cent a year, *i.e.* about the average for 1900–1938.

POPULOUS CENTRE OF WORLD CAPITALISM

The most populous centre of world capitalism has always been and still is Western Europe. At the middle of the century rather more than half of the total population of capitalist countries lived there, but the growth rate in most West European countries has been very low in recent decades compared with other geographical regions, while emigration has been very high, which has deepened the trend already long apparent towards a reduction of the weight of Western Europe's population in both the capitalist countries as a group and the capitalist world as a whole.

The population of the North American area increased much faster than Western Europe's after the war. In the period 1950–81 its annual average growth rate was 1.30 per cent, which led to a certain convergence of the absolute size of the populations of the two main centres of capitalism. While the ratio between them was 1.7 : 1 in favour of Western Europe in the early 50s, it was 1.4 : 1 at the end of the 70s.

p One must remember, however, that the population of North America was increased in the first half of the century, and especially as a result of World War II, by a flow of immigrants from outside, including Western Europe, in addition to a rate of natural growth higher than in other developed capitalist countries. At the same time there was a high population growth ratc in other capitalist countries, above all in the biggest of them, Japan, although it was also much lower than in the developing countries. A consequence of all these factors was a considerable decline in the weight of the industrial centres in the total population of today's capitalist world. The population of Africa (not counting South Africa) rose by 210 million, or more than doubled, in the same period, and its average annual growth rate began to exceed that of Asia in the 70s. These rates were not, however, limiting ones. Latin America set records of sorts, its population increasing by 120 per cent by the end of the 70s compared with the early 50s, with a mean annual growth rate of 2.75 per cent.

These major shifts in the ratio of population among the various groups of countries convincingly emphasise the need to make careful allowance for the demographic factors when studying the long-term development trends of the productive forces in the postwar capitalist economy. And they are of special significance for a sectoral analysis of the results of the main regions' economic development. At the same time these factors are extremely important for forecasting growth of the productive forces in any one economic system and on a global scale. The Soviet sociologist Arab-Ogly has justifiably commented that the tempestuous growth of population is one of the most important historical phenomena of our time as regards its remote consequences. It is facing both individual countries and mankind as a whole with a number of exceptionally complicated though in principle resolvable economic, social, and political problems. That is why there is an urgent need for substantiated and as exact as possible forecasts down to the year 2000. The compilation of such forecasts has now acquired universally acknowledged theoretical and practical significance. Demographic processes as a rule have immense inertia, which also makes it possible, strictly speaking, to make more or less substantiated forecasts of future populations.

In spite of the marked rise in tempos in various countries and years, the dynamics of population growth is limited in the principal regions by quite definite parameters, which provides the necessary premises for a more or less long-term estimate of the future distribution of manpower among them. A real effort has been made in UN demographic publications to tie up population growth trends somehow with certain outlooks for the subsequent course of development of production. One of its surveys remarked, for instance: The future size, structure and distribution of population are essential for any plan that involves food, housing, employment' education, health or other public services. This is also an era of increasing awareness: of man's unprecedented growth, of the interaction between this growth and the environment, and of the possible implications for the future. Investigations into these complex relationships have added to the demand for population projections.

For all the substantial differences in estimates of the future development of the international demographic situation, the majority of them quite validly, in our view, assume that one can hardly expect any extraordinary changes in growth rates in the really foreseeable term either in the capitalist or the developing countries compared with the postwar period, though some demographers' forecasts assume a certain lowering of these rates in coming years. In any case, however, there are adequate grounds for staling that the number of inhabitants of all the countries of the non-socialist world will increase by at least 50 per cent in the last two decades of the century (16 to 20 per cent in the first group and 56 to 60 per cent in the second). In absolute terms this means an increase in the population of the industrialised countries of roughly 120 to 150 million, and in the developing countries of 1,200 to 1,300 million.

These estimates, like all similar generalising forecasts, are approximate, but in themselves they are very important not simply for understanding the outlooks for the development of population but even more for clarifying their interconnections with the possible trends of the future growth and distribution of the productive forces in the countries at present forming the world capilalisl system. The postwar period has been marked by a further increase in the unevenness of the distribution of the main productive force in the developing world. Most of its inhabitants, and so of its manpower, has long been concentrated, as we know, in Asia, whose weight in the total population of the capitalist world has been steadily rising in postwar decades. Analysis of current trends gives grounds for sugg-esting that the Asiatic population will be around half in the mid-80s, and more than half in Ihe 90?, of the total population of the capitalist world within its present boundaries. But the majority of the liberated countries of Asia (in which more than 2,000 million people will probably be living at the end of the century, *i.e.* three times as many as in the middle of the century) are now the most backward areas, economically speaking, of the developing world. The population of Africa and Latin America is increasing even faster. These expected shifts arc undoubtedly fraught with serious socio-economic consequences for the fate of capitalism in those areas.

In coming decades jobs will have to be created in them for hundreds and hundreds of millions of people, but their agriculture is already unable to absorb any considerable part of the flood of new labour power. The objectively operating laws of agricultural production, moreover, are leading to surplus manpower being pushed out into the towns which, given capitalist relations of production, inevitably means a growth of poverty and proletarianisation of broad masses of the peasantry. That is one of the important distinguishing features of capitalism compared with the exploiter formations that preceded it. The law of faster growth of the urban population under the capitalist mode of production, Lenin stressed, 'does not and cannot operate otherwise than through the disintegration of the peasantry into a bourgeoisie and a proletariat'.

When that conclusion was drawn there was still no socialist economic system. Experience of its development has shown that for it advancc of society's productive forces is inseparably linked with priority growth of the urban population, but it also indicates that the new mode of production, unlike the capitalist, makes it possible to avoid the extremely grave consequences of urbanisation for the broad masses objectively inevitable for all capitalist countries.

Since the war urbanisation has intensified to an unusual degree under the impact of the shifts in the structure of social production considered above. It has begun to have an ever more active effect on the character of the distribution of manpower in the world capitalist economy. In the 50s, 60s, and 70s the urban population of non-socialist countries more than doubled, so that the ratio of urban and rural population altered, with socio-economic and political consequences of no little importance. At the end of the 70s more than 40 per cent of the capitalist world's total population was already living in towns, while less than a third had done so in the early 50s. In the industrial countries this proportion rose to 75 per cent, and in the developing countries to a third, with fewer and fewer of the rural inhabitants available for work being directly involved in agrarian production. At the same time an increasingly large amount of manpower is becoming concentrated in the towns, a trend that has become stable not only in industrially developed capitalist countries but also in the overwhelming majority of developing countries.

There are, of course, considerable regional differences in this respect in both groups, largely associated with features of their preceding historical development. In the industrial centres the highest proportion of urban population is in North America, particularly in the United States, where more than 90 per cent of the total population was living in towns at the end of the 70s. There is an even more significant spread in the weight of the urban population in developing countries, from 15 or 20 per cent in Asia to 60 or 65 per cent in Latin America. One of the most characteristic trends in recent urbanisation is the rapid increase in the role of former colonial regions in the total urban population of the capitalist world in both absolute terms and relatively. At the beginning of the period reviewed the bulk of the urban population lived in industrial countries, while at the end already more than half of the townsmen lived in the agrarian and primary producing countries. The size of this population grew roughly twice as fast in the latter as in the former, and analysis of the long-term trend allows us to suggest that more than half of the population of the contemporary capitalist will be townsmen by the end of the century. At least twothirds of them will be living in the former colonies and semi-colonies of Asia, Africa, and Latin America.

A not unimportant feature of the long-term shifts in the distribution of population is its growing concentration in very big cities. The number of cities with more than a million inhabitants increased in the 50s and 60s alone by more than 50 per cent, while the number of their inhabitants more than doubled. Big cities have been growing much faster, moreover, in developing countries. Most of the biggest cities of the contemporary capitalist, world are now already located in developing countries. The chaos of capitalist urbanisation is leading to the formation of broad urbanised zones with multimillion populations in both groups of countries, which are drawing an ever bigger part of the population of adjoining towns and rural areas into their orbit, thereby intensifying the unevenness inherent in capitalism and anarchy in the distribution of society's main productive force.

Substantial shifts are also taking place in the geographical distribution of the rural population of the non-socialist world, which has grown in recent decades exclusively in the agrarian and primary commodity producing countries. In the industrially developed countries there has been an absolute fall in its numbers, while it has increased by more than two-thirds in the developing countries as a whole, and exceeded 1,600 million at the end of the 70s. In the principal economic centres the complete ousting of the petty producer from agriculture was completed, or almost so, in the postwar decades. Engels pointed out the objective inevitability of this in his time, remarking that the development of the capitalist form of production has cut the lifestrings of small production in agriculture; small production is irretrievably going to rack and ruin.

AMOUNT OF WORLD CAPITALISM'S RURAL POPULATION

In the majority of emancipated countries that have taken the road of capitalist agrarian relations, this process is developing with increasing force, but attempts to force its pace are still complicated by the low level of development of the other spheres of their economies, above all of industry. The proportion of the industrial countries in the total rural population consequently fell by half in the 50s through 70s, while it rose from less than 75 per cent in the developing countries to 90 per cent. As a result there was a considerable extension of agrarian overpopulation in the latter. This enormous overpopulation, which was long a heavy burden on their economies, still remains one of the most disastrous social 'side-effects' accompanying their economic growth under capitalism. An increasing amount of world capitalism's rural population is thus concentrated in developing countries, above all in Asia and Africa, where the rapidly growing towns are unable to absorb the villages' immense surplus of manpower. At the beginning of the 80s around two-thirds of this rural population lived in Asia and more than 20 per cent in Africa. The only major region where its proportion has begun to fall is Latin America, and there too, in fact, it still maintains its tendency to grow in absolute terms (increasing by more than 40 per cent in the period under review). In the coming decades the marked imbalance will apparently not be smoothed out, but will become rather more acute. In fact, all in any way valid long-term demographic forecasts envisage a progressive reduction of the rural population in capitalist countries and an inevitable growth of it in developing ones by at least a third by the end of the century.

The absence of sufficiently reliable estimates of the agrarian overpopulation in developing countries is a factor making it extremely difficult to determine exactly the scale and dynamics of employment in the whole world capitalist economy. The inadequacy, in turn, of the data on full or partial unemployment in industry and other spheres of production in most developing countries, and in some capitalist ones, is also an obstacle; so, too, are the constant upswings and recessions in employment through the effect of marked cyclic and non-cyclic fluctuations of the capitalist economy. Still, however, there arc composite indicators in the international statistics that enahle us to estimate, with the necessary degree of comparability, the growth trends in the long term of that part oi' the population that could be involved in the reproduction process. This type of indicator primarily includes estimates of the scale of growth of the gainfully employed population, analysis of which helps us arrive at the characteristics of the development trends of productivity for the different groups of countries.

The imbalance in the postwar growth of the able-bodied population between the two groups of the capitalist economy has increased immensely. Over the twenty years of the 60s and 70s, for example, their numbers increased twice as slowly in the developed capitalist countries as in the developing ones. Consequently there was a tendency for the proportion of the former in the total gainfully employed population of the capitalist world to decline. From the fact that the decisive mass of the population that will begin to work at that time in both groups of countries already exists, we can assume that around three-quarters of the gainfully employed population of today's nonsocialist world will be concentrated in the now developing countries of Asia, Africa, and Latin America in the mid-80s, and around i'our-iifths by the end of the century. But this labour force will still obviously be relatively unskilled in the main while the most skilled part will be concentrated as before in the economically developed centres.

In that connection the problem of the migration of various categories of workers between countries with different levels of development will be aggravated. In many industrial countries there has been a fast growing shortage since the war of national personnel in the mass trades that do not call for any degree of professional skill. The shortage began to be covered more and more by importing labour from economically less developed countries. In the 70s this shortage constituted around ten or twelve million, and may (according to some estimates) increase several times over again by the end of the century. As for the developing countries the inherent tasks of their socio-economic development will steadily lead, in the first place, to an increase in their need for well-trained, skilled personnel, a need that cannot be wholly met simply from local resources. The possibilities for attracting specialists from outside, however, are greatly limited by the increasing need for such personnel in the developed countries themselves, and also by the comparatively high standard of their pay there. It is not fortuitous that a mounting portion of the West's funds for 'technical aid' are spent on the specialists sent to developing countries, and on training students.

At the same time, monopoly capital, utilising the laws of capitalist competition, is everywhere broadening the 'import' from developing countries of the most talented and capable part of their intelligentsia and skilled specialists. The 'brain drain' from Asia, Africa, and Latin America has become a marked feature of the international migration of labour. The spontaneous flow of surplus labour between separate developing countries has also become unparalleled in scale in recent decades.

INTERNATIONALISALION OF SOCIAL PRODUCTION

All these objective 'costs' of the internationalisalion of social production are making world capitalism's demographic problems extremely complicated and increasing their tension. At the present level of development of mankind's productive forces the registering, regulation, and planning of manpower are already becoming more and more urgent on both the national and the international level, but capitalist society, rent by antagonistic contradictions, is unable in practice to face up to dealing with them. They can only be coped with by a more rationally structured progressive social system, a thought, that Engels expressed as follows: If it should become necessary for communist society to regulate the production of men, just as it will have already regulated the production of things, then it, and it alone, will bo able to do this without difficulties.

In most areas of the capitalist world there has been a tendency since the war towards a certain reduction of the proportion of the gainfully employed population in the total population. The operation of this trend is linked in the main with the considerable changes in the age structure of the population, especially in the developing countries, where there has been a marked extension of the age brackets of people who have not yet reached maturity. In many capitalist countries, above in Western Europe, another factor underlies the lowering of the proportion of the ablebodied, namely the rise in the expectation of life and increase in the age bracket of elderly people of pensionable age. Analysis of the age factors leads to the conclusion that an increasing part of the most active age brackets will probably be concentrated in coming years in the developing countries, from which there follow very complex demographic, and consequently socio-economic, consequences for further shifts in the distribution of the main productive force within the world capitalist economy. The indices of the dynamics and scale of the distribution of manpower in the two groups of countries will become even more significant when they are compiled from the data cited above on growth of the aggregate product. This comparison makes it possible to bring out the long-term trends in growth of the social productivity of labour calculated on the labour force potentially existing in the non-socialist world, *i.e.* on the size of the gainfully employed population. This productivity can be expressed most generally as the difference between the mean annual growth rates of the GDP in every macro-economic system (or its sub-systems) and the corresponding indicators of the growth of manpower.

Because of their generalised character the results arrived at are very approximate and require further elaboration and concretisation both for the groups of countries and their leading industries and for the various periods. It follows from them that, with all sorts of fluctuations of production and employment in the two groups of countries, the mean annual growth rate of potential productivity of social labour, calculated from the dynamics of the gainfully employed population over the whole period considered was around 3.1 per cent in the centres of capitalism and 3.2 per cent in the developing countries. This relation was extremely unstable, however, and was by no means typical for all the postwar decades.

In the first years after the war the considerably accelerated economic development in the main agrarian economic regions, and the comparatively low rates of manpower growth preserved in them from the colonial epoch, led to a tendency towards accelerated growth of their extremely low productivity of social labour, reaching the average indices of the main centres of capitalism in the 50s. In the 60s, however, there was again a revival of capitalism's inherent line towards widening of the gap in this respect belwcen its centres and periphery, which was furthered by the industrial countries' active taking up in those years of the advances of the scientific and technical revolution, though within the system as a whole insuperable inner contradictions fraught with new economic upheavals continued to accumulate.

The extremely acute crisis processes that developed in the rnid-70s in the world capitalist economy were marked by an unprecedented weakening of the economic effectiveness of exploiting the advances of this revolution in practice in all Ihe industrially developed countries. The substantial slowing of growth and later the absolute fall in production of their aggregate GDP could not help being reflected as well in the productivity of social labour. The major postwar economic crisis of the mid-70s laid bare the growing instability of the world capitalist system and the extremely contradictory character of the effect of modern scientific and technical progress on it as regards growth of society's productive forces. It would hardly be proper, however, to say that the objective patterns leading to the creation of such a vast imbalance in the levels of the social productivity of labour between the two groups of countries at the different poles of this system will begin to lose their former significance in coming years. In all probability this imbalance and steep fluctuations of its dynamics will remain essential features of the world capitalist economy's further development.

The processes mentioned indicate simply certain very general resultant trends in the postwar shifts in the economies of capitalist and developing countries. Within each of these groups there is in turn a very wide range in the levels, dynamics, and conditions of the growth of social productivity. Above all we must draw attention to the considerable strengthening of unevenness in the distribution of the productive forces in the developed capitalist countries in recent decades.

INDUSTRIAL CAPITALS FOR PRODUCTION

The external relations variable pertained to level of dependence on industrial capitals for production inputs and purchase of farm outputs and to the degree of involvement in credit relations, especially with finance capitals. Applying this theoretically based typology to the data from three areas in southern England — an urban fringe area, an agricultural area undergoing commercial and industrial development, and a primarily agricultural area — they found that the process of subsumption of farm production relations, and thus their integration into the "wider circuits of capital," varied over time and space. Whatmore et al. concluded that farm families need to be understood as actors in the process of subsumption rather than its passive victims.

The neo-Marxian theoretical enterprise (and that of the modernization school as well) on the conversion of traditional peasant economic relations to commodity relations under capitalism has also been scrutinized by Vandergeest. Among his major criticisms were that neo-Marxist approaches tend to be too unilinear, too macro-structural, too denying of human agency, too inattentive to the role of the state in the process, and too unconnected with practice.

He has argued, from a neo-Weberian position rooted in the work of Pierre Bourdieu, that all categories and theories are in the last analysis historically contingent, ideological, and interpretive social products.

The world is a complex whole which can only be investigated empirically and understood through theory. It cannot be reduced to the simple working out of a model derived through deductive theory — whether that of "simple commodity production" or of "capitalism". There is not a single deductive logic (such as the logic of exchange-value or accumulation) underlying or determining all relations in a capitalist formation, but there are different, historically contingent principles which we can only investigate through empirical research.

There has recently been a tendency in the neo-Marxist and neo-Weberian literatures towards a deemphasis on "deductivist"-functionalist theoretical postures. Subjectivist perspectives such as those of Vandergeest and Mooney have been influential in leading to this reorientation, though this tendency is by no means limited to those persuaded by neo-Weberianism. The three postures in the Marxist political economy of agricultural tradition — those emphasizing the persistence of petty commodity production, those emphasizing the inevitable differentiation of petty commodity producers into antagonistic social classes, and those seeking synthetic positions — have yielded a continuing, lively debate in the rural sociology literature. Particularly instructive is the debate over the Mann-Dickinson hypothesis.

Also of importance is Goodman and Redclift's commentary on the work of Friedmann in which Goodman and Redclift critique Friedmann for taking an overly deductive approach to petty commodity production that ignores the role of historical conjuncture and ideology. Bernstein has made a comparable argument — that generic theories of simple commodity production under advanced capitalism tend to overemphasize its "functional" aspects and downplay contradictions and the diversity of household forms of production.

It should be emphasized that theoretical and empirical treatments of the political economy of North American agriculture have not been made only by rural sociologists. Many of the pioneers in this tradition of scholarship have been nonsociologists or sociologists who have had little or no connection with the Rural Sociological Society or the institutionalized form of rural sociology in U.S. land-grant universities. Nonetheless, rural sociologists have made influential contributions to this literature.

The emerging political economy tradition in the sociology of agriculture has led to a provocative and stimulating literature. It should be mentioned, however, that those working from this perspective are just beginning to establish a distinctive research programme; much of the existing literature, in fact, often tends to involve superimposing a new vocabulary on already-established data, such as Goss et al. set out to do. Establishment of a distinctive research tradition will probably occur slowly since conventional data sources such as censuses and sample surveys are often inapplicable to theoretical issues in the political economy of agriculture. There have, nonetheless, been several research papers in the political economy tradition that have utilized innovative new methodologies to generate research findings that bear directly on specific hypotheses.

SUBCULTURE AND AGRICULTURAL STRUCTURE

It is useful to note that the growing interest in subjectivist approaches in the new sociology of agriculture has not been confined to those who work within the neo-Marxist and neoWeberian traditions. Anthropologically-oriented researchers have also contributed to this literature through the examination of the effects of ethnic background on state and local level farm structure. Salamon and associates have done the pathbreaking studies in the anthropological tradition in Illinois communities, though recently this work has been taken up by others. While their work was not intended to contribute to the "structure versus agency" debate that has raged between Mann and Dickinson on one hand, and Mooney on the other, and actually predated much of this debate, it is clearly relevant to the debate. In particular, the work of Salamon and associates pertains to the role of subjective motivations in farm decision making, especially motivations conditioned by persisting subcultural variations based in differences in ethnic origins of farm families.

The key finding of this line of research has been that farm families with German ethnic backgrounds tend to view farming as a way of life and hold a strong value for keeping the family farm intact. These Salamon has labeled "yeoman" farmers. In contrast, farm families with British ethnic backgrounds tend to be more entrepreneurially oriented, viewing farming as a way to make profits and having little attachment to farming or to particular farms. These Yankee farmers are more likely to seek growth in scale and less likely to support their local communities. This research has led to understanding differentials in farm size as a partial function of differences between these subcultural variants. Flora and Stitz found that yeoman farmers in a Kansas county generally expanded less than did Yankee farmers. Foster et al. also found support for the Salamon hypothesis in a study of a sample of Illinois farms that had been in the same families for 100 years or more.

Several other anthropological studies in the tradition of Salamon and colleagues have been reported in Chibnik. Of particular importance is Barlett's article in which she criticizes notions of the demise of the family farm and of the "disappearing middle" from an anthropological perspective.

SOCIALIST MODE OF PRODUCTION

After World War I there was a comparatively rapid restoration and further expansion of production, and of the world economic links broken by the war, within contracted boundaries. But this growth took place on a background of further deepening of imperialism's basic contradictions, a sapping of its colonial foundations, and an intensification of the unevenness in the development of separate regions and countries.

At the same time the socialist mode of production was more and more clearly becoming a constituent part of the global economy of mankind, and was graphically demonstrating its vitality.

The international standing of the Soviet Union was strengthened, and its external economic 51 relations extended, through intergovernmental agreements with several capitalist conn tries, certain former colonies and semi-colonies. The progress made in building socialism in the USSR, and the consolidation of its position on the international arena, where the foundations of a new world social system were being laid, became vital factors in the subsequent aggravation of the general crisis of capitalism.

Growth of capitalism's irreconcilable contradictions continued to play the determinant role in this process. Particularly convincing evidence of that was the world cyclical overproduction crisis of 1929–33, the greatest in scope, duration, and disruptive consequences in capitalism's history, which simultaneously swept the industrially developed countries and their agrarian, raw material periphery. The crisis, which threw the economy of the capitalist world several years back, had a very negative effect as well on the whole system of imperialism's global economic connections. By the end of the 30s the volume of international trade, and also of capitalist industrial production was below the level of 1913. Right up to the beginning of World War the economies of most countries of the capitalist world had still not been able to recover fully from that crisis. At the same time the colonial foundations of the capitalist economy had- been markedly weakened. The successful anti-imperialist struggle of the peoples of the USSR and their achievements on the road of radical socio-economic reforms, stimulated an upsurge of the national liberation movement in many colonies and dependent countries, some of which had managed after World War 1 to win or retain their national independence. A crisis of the all-embracing colonial system began. The colonial powers, it is true, were then still able to avert the break-up of their empires. Without ceasing their struggle to redivide spheres of domination, they endeavoured as before to deprive the countries economically dependent on them of national sovereignty; and to that end imperialism repeatedly unleashed predatory wars (*e.g.* Italy's aggression against Ethiopia, and Japan's against China).

Characteristically, however, not a single imperialist power could succeed, even in the initial stage of the general crisis of capitalism, in turning any economically backward country into a new colony. While retaining a decisive position in the world economy, linance capital was forced to take into account in its expansionist policy not only the 'strength' and 'capital' of its imperialist rivals but also of the growing strength of resistance of the peoples of the countries exploited by them, who received every possible support from the iirst socialist country. World War II, which broke out as a result of the sharp aggravation of all capitalism's contradictions at the end of the 30s, and of the striving of its most aggressive circles to strike a crushing blow against socialism and the national liberation movement, in the final count led to a further deepening of the crisis of the imperialist system, the boundaries of which contracted even further. The socialist mode of production became predominant in a number of countries in Europe and Asia, and a new phase opened in the economic competition of socialism and capitalism.

The antagonism and interaction of the two opposing social systems became a determining feature of today's world economy in the postwar years; in that connection it has become necessary, when studying the structural shifts and fundamental features of the postwar development of capitalism, to take into account at the same time the global tendencies of the growth of production, the more so that the long-term consequences of the scientific and technical revolution that has embraced the leading spheres of the productive activity of human society as a whole are beginning to have a mounting effect on both socialist and capitalist countries, and also on many developing ones. Analysis of the facts characterising these tendencies indicates that in the period under review mankind's productive forces have attained a height hitherto unprecedented. The international exchange of scientific and technical knowledge and production know-how has reached a fundamentally new level. The infrastructure of world economic relations has also developed rapidly under the impact of the scientific and technical revolution on the basis of a national and international division of labour. On the whole, the volume of the aggregate product of all the countries of the world was more than four times as big at the end of the 70s as at the beginning of the 50s. There was also a marked increase in the main productive forces, *viz.*, manpower (the 53 number of the world's inhabitants rose by almost 75 per cen! in the same period).

The list of that kind of quantitative indicators could, of course, be considerably extended, but the figures cited give a graphic idea of the sizable shifts that have occurred in humanity's productive forces in the contemporary historical situation of competition of the two main modes of production and the corresponding two world socio-economic systems, namely the socialist and the capitalist. As Karl Marx remarked in a letter to P. V. Annenkov (28 December 1846), tho productive forces are therefore the result of practically applied human energy; but this energy is itself conditioned by the circumstances in which men find themselves. Which of the opposing world social systems has the advantage in creating conditions for growth of the productive forces can be judged from the following facts. On the whole, in the three decades of the 50s through to 80s, the rates of growth of national income and industrial output in the socialist countries belonging to the Council for Mutual Economic Assistance (CMEA) were three time as high as *in* all the capitalistically developed countries taken together. The countries belonging to CMEA are the most dynamically developing economic community in the world. At the end of the 70s these countries, which have 10.2 per cent of tho world population, produced around one-third of the world's industrial output. In 1968–79 alone the volume of the gross social product in the European members of CMEA increased more than fourfold, while that of all the developed capitalist countries increased by only 140 per cent. A natural consequence of the development of this tendency has been a considerable raising of the role of socialism in the world economy. At the beginning of 1980 the population of 54 countries not yet broken free from direct, colonial dependence was not more than 0.1 per cent of all mankind.

The turn of the decade to the 80s was marked by further growth of the economic might of the countries of world socialism. As the Central Committee's report to the 26th Congress of the Communist Party of the Soviet Union noted: The past few years have not, been among the most favourable for the national economies of some socialist states. Still, in the past ten years the economic growth rates of the CMEA countries have boen twice those of the developed capitalist countries. Tho CMEA members continued to bo the most dynamically developing group of countries in the world.

At the same time the break-up of the colonial system was not limited to emancipation of the colonies, but also included semi-colonial and dependent countries. Throughout the long period of'the imperialist powers' unlimited sway in the world economy the dominant tendency was to convert the latter into rightless colonies; in the postwar years another trend began to predominate, that of converting them into politically independent national states. Imperialism could not suspend the development of this tendency as well. In recent decades the overwhelming majority of semi-colonies and dependent countries, having thrown off foreign and proimperialist regimes, have also in fact broken free of imperialism's system politically, and are no longer its reserve, as they used to be. With the all-round support of the socialist countries and international working-class movement, they are intensifying the anti-imperialist struggle for economic independence. As a result there has been a marked quickening of the growth rates of their economies. The total production of goods and services in the developing countries of Asia, Africa, and Latin America rose more than 320 per cent in the 50s to 80s, industrial output rising by 620 per cent and farm production by nearly 130 per cent.

Their position in the world capitalist economy, however, remains extremely difficult and unequal. In spite of the fact that they have three-quarters of the population of the non-socialist world, they produced less than one-fifth of 55 that world's gross output at the beginning of the 80s. The immense gap in levels of development between them and the group of imperialist powers historically built up under capitalism has continued to widen. In the first postwar years, for example, this gap, in terms of per capita GDP, was characterised by a ratio of 1: 10, and at the beginning of the 80s by a ratio of more than 1: 13. In the same period there was a marked reduction of the weight of the former colonial world in the turnover of the world capitalist market. The capitalist economic system is more and more clearly demonstrating its incapacity to deal with the pressing socioeconomic problems of emancipated countries. The logic of history is convincing their peoples of the lack of perspective of development along the capitalist road.

It is not fortuitous that many of the new national states, including some that are still largely enmeshed in imperialism's world economic relations, are rejecting the capitalist road of development in their constitutions, under pressure of the broad public, and are proclaiming a line of building a socialist society in the long term. Leonid Brezhnev, characterising the significance of the countries liberated from colonial subjection that have taken the road of revolutionary, democratic transformations, and of countries with a socialist orientation in today's world, said that at the 26th Congress of the CPSU:

Development along the progressive road is not, of course, the same from country to country, and proceeds in difficult conditions. But tho main lines are *similar*. These include gradual elimination of the positions of imperialist monopoly, of the local 56 big bourgeoisie and the feudal elements, and restriction of foreign capital. They include the securing by the people's state of commanding heights in the economy and transition to planned development of the productive forces, and encouragement of the cooperative movement in the countryside. They include enhancing the role of the working masses in social life, and gradually reinforcing the state apparatus with national personnel faithful to the people. They include anti-imperialist foreign policy. Revolutionary parties expressing the interests of the broad mass of the working people are growing stronger there.

MONOPOLY CAPITALISM

Monopoly capitalism is really losing its old opportunity to decide the fate of nations even within the shrunken framework of the world capitalist economy. The main point in the people's national liberation struggle at the present stage of the general crisis of capitalism"is that it has been converted in practice into a fight against exploiter relations in general, both feudal and capitalist, in many countries. As Leonid Brezhnev has put it:

It is of exceptional importance that many of the countries that have achieved liberation have rejected the capitalist road of development and adopted a socialist orientation, setting themselves the goal of building a society free from exploitation. And that, without doubt, is a very strong blow to capitalism's positions as a whole as a world social system. A substantial socio-economic consequence of the collapse of the colonial system is the relative lowering of the role of expatriate capital in the economies of emancipated countries. Before independence almost all their industrial production was in the hands of foreign monopolies and was part of the economy of the imperialist system. Today broad measures have been taken in'many of them both to nationalise expatriate enterprises and establish public control over their activity, and to found their own national industry. There are few summarised, reliable facts in the international statistics on what proportion of industry remains under the direct control of expatriate monopolies, but varrious indirect indicators suggest that a considerable^part of the industrial potential of the developing countries, yielding more than 4 per cent "of gross world industrial output, has already passed *out* of such control.

Because of the effect of these tendencies, a further weakening of the imperialist powers' positions in the world economy has become typical of the period of the general crisis of capitalism, which finds expression above all in the obvious undermining of their former hegemony in world industrial production. At the end of the 30s the monopolies of the capitalist countries (including colonies and dependent countries) in practice controlled around nine-tenths of total world industrial production; at the beginning of the 80s only a little more than half of it was under their control. The facts adduced here primarily bring out one of the decisive trends in the crisis of the imperialist economy, when the basic features and main peculiarities of contemporary world development are beginning to be determined by the working class and its offspring, the world socialist system, and no longer by the monopoly capitalists. The steady strengthening of the economic might of the socialist countries, the advance of the national economies of the new sovereign states, and the rapid development of mutually advantageous international economic relations between them are leading in turn to a marked limitation of the imperialist monopolies' diktat in world economic relations, including those between developing and developed capitalist countries on the world market.

Even the most persistent and subtle defenders and propagandists of the capitalist system are unable to deny the tendencies, which are finding wide reflection in international statistics, and which express the main trends in the deepening of the contemporary crisis of the world capitalist economy. An ambivalent approach to these processes is therefore typical of capitalist politicians, economists, and sociologists, who are engaged in apologetics for imperialism. On the one hand, they prefer in every way to evade comparison of the long-term results of the development of the two world economic systems in their interconnections, and strive to avoid a comprehensive socio-economic analysis of these results. While the work of some leading capitalist scholars 58 undertakes to examine the dynamics of 14io economic growth of the opposing social systems today, it usually mentions the advances of socialism in passing as something irrelevant for an evaluation of the present-day level and the ensuing outlook for the global development of mankind's productive forces. On the other hand, immense attention is paid in Western capitalist literature to another aspect of this problem, namely, the factors that continue to promote growth of production within the capitalist economy. These factors are usually considered out of context of the processes showing overtaking growth rates in the economy of the world socialist system.

The ideologists of the capitalist system, glossing over its class and other antagonisms in every way, are relying on the development trends of social development in its industrial centres under the impact of the developing scientific and technical revolution. What lack of prospects can there he talk of, they ask, and even more what hopeless crisis of this social system, if capitalism's productive forces still continue to grow, albeit unstably and with sharp fluctuations. The increase in the production capacity of the postwar world capitalist economy is thus advanced as the 'weightiest' argument to substantiate the 'historical stability' of capitalism and consequently, as well, the flimsiness of the Marxist-Leninist analysis of the irreconcilable contradictions of its development. In fact the Marxist-Leninist theory disclosed another causal connection between the growth and internationalisation of social production under capitalism and its historical destinies, than that which its capitalist critics try to ascribe to it. The objective tendencies in the development of the capitalist mode of production are not leading to a stopping of growth of the productive forces, or to their stagnation, but on the contrary are leading to their accelerated advance compared with preceding formations. In the *Manifesto of the Communist Party* Marx and Engels wrote:

The bourgeoisie cannot exist without constantly revolutionising the instruments of production, and thereby the relations of production, and with them the whole relations of society—Constant revolutionising of production, uninterrupted disturbance of all social conditions, everlasting uncertainty and agitation distinguish tho bourgeois epoch i'rom all earlier ones.

But does all that mean that, as the productive forces of capitalist society develop, its economic basis is perpetuated and becomes ever firmer and more unshakeable? The founders of Marxism-Leninism, in bringing out the inner contradictory dialectic of that development, demonstrated that it in no way contributes in the end to the consolidation of capitalism. It relentlessly undermines the foundations of this system, and forms the conditions within it necessary, for victory of a more advanced social system. By accelerating growth of the productive forces in its own class interests and internationalising them in every way, the capitalist class itself prepares the material and social conditions for destruction of the social formation created by it.

The transition to monopoly capitalism, while heightening the unevenness and spasmodic character of its development, was marked by a further growth of capitalist production. In describing imperialism as the highest and last stage of capitalism, in which tendencies towards decay and growth of society's productive forces predominated simultaneously, Lenin stressed that:

It would bo a mistake to believe that this tendency to decay precludes the rapid growth of capitalism. It does not. In the epoch of imperialism, certain branches of industry, certain strata of tho bourgeoisie and certain countries betray, to a greater or lessor degree, now one and now another of these tendencies. On tho whole, capitalism is growing far more rapidly than before; but this growth is not only becoming more uneven in general, its unovenness also manifests itself, in particular, in tho decay of tho countries which are richest in capital.

The very substantial shifts in capitalism's material basis in its imperialist stage graphically confirm the fundamental character of these tenets of Marxist-Leninist theory, which help us better to understand the action of the long-term trends in the world economy that predetermine the features and course of its development in today's transitional period. As the general crisis of capitalism deepens, the world capitalist economy more and more sharply experiences the effect of the socialist and national liberation revolutions, the mounting militarisation of the economy, and of numerous military shocks and upheavals. The development of 60 production in all the countries of this economy has been repeatedly interrupted or held back by economic crises, but the objectively operating long-term tendency towards growth of society's productive forces has still continued to make itself felt in the stage of imperialism as well. The economies of capitalist countries began to develop at particularly high rates after World War II, as witness the following facts. Fundamentally new branches of industry, transport, and communications were developed then under the impact of the scientific and technical revolution. Internationalisation of social production was noticeably broadened and deepened; the physical volume of international trade on the world capitalist market rose nearly eightfold.

The number of working people, and their vanguard the working class, increased on an unprecedented scale in the postwar decades. When Marx and Engels disclosed the very great revolutionary potential of the proletariat in the *Manifesto of the Communist Party*, there were only around nine million workers in the whole world. At the beginning of the 1980s their numbers had increased nearly 50-fold. The working class has become the leader of the broad masses' light for a revolutionary restructuring of society. And only in that struggle could the new social system arise that is going to replace capitalism. Capitalism, Lenin said, creates its own grave-digger, itself creates the elements of a new system, yet, at the same time, without a 'leap' these individual elements change nothing in ths general state of affairs and do not affect the rule of capital. The patterns of the general crisis are not introduced into the world capitalist system from outside but are the natural result of its inner development and stem from the antagonistic nature of the capitalist mode of production itself, based on the exploitation of man by man and of some countries by others. The decisive consequence of the structural shifts under review was not a weakening but an unprecedented sharpening of the basic contradiction of capitalism, *i.e.* the contradiction between the constantly increasing social character of production and the capitalist, exploiter character of the appropriation of social labour. It is on that basis that all the other contradictions of modern capitalism and its world economy continue to develop. A thorough study of the capitalist economy of his time enabled Karl Marx to conclude that the historical development of the antagonisms, immanent in a given form of production, is the only way in which that form of production can be dissolved and a new form established.

Lenin repeatedly stressed the paramount importance of extending this conclusion of Marx's to analysis of the monopoly stage of capitalism, and pointed out, already at the beginning of the contemporary transitional epoch that: It would be impossible to put an end to the rule of capitalism if the whole course of economic development in the capitalist countries did not lead up to it. Capitalist economics lias repeatedly tried in this connection to give battle to Marxist-Leninist political economy on the problems of world economic crises, in which, as Engels put it, 'the contradiction between socialised production and capitalist appropriation ends in a violent explosion'.

CAPITALIST STATISTICS

Many capitalist economists have tried with redoubled energy in the postwar period to demonstrate that capitalism would manage to enter a fundamentally new phase of development of the productive forces by means of new methods of economic policy and state-monopoly control, and that 63 the former instability and profound contradictions of the cyclic development of its economy would largely lose their force under the action of the capitalist state. The very facts of capitalist statistics, however, prove that since World War II, and throughout the preceding history of monopoly capitalism, not a single capitalist country has been able to avoid considerable cyclic fluctuations and slumps in its economy. Periods of relatively high growth rates have inevitably alternated with periods of low ones, and sometimes even with an absolute fall in production and trade. The gap between the highest and lowest mean annual growth rates of the gross product was almost fivefold on the whole in the 50s and 60s, and in world industrial production and international trade it was even greater (from—2.5 per cent to -\-*Q.5* per cent and from—3 per cent to -\-*13* per cent respectively). The three economic slumps suffered by the world capitalist economy in the 70s and early 80s, including that of 1974–75, the deepest since the war, as was brought out iu detail at the 26th Congress of the CPSU, confirmed with all obviousness the irrefutable fact that state-monopoly control of the economy has proved incapable of curbing capitalism's elemental forces. The new phenomena include an immense extension of the role of international monopolies in the economy of modern capitalism. It is their external economic expansion, unprecedented in scale and consequences, that lias largely determined the postwar features of the development of the world capitalist economy considered above. This expansion has acquired such a scale that the leading organs of the capitalist press and international economic organisations write about it with growing alarm. The special study *Multinational Corporations in World Development* made by UN experts said in particular:

In the past quarter of a century the world has witnessed the dramatic development of the multinational corporation into a major phenomenon in international economic relations.

The international monopolies, concentrating immense economic might in their hands, had gained control in the 70s of more than a liftli hotli of the production and the distribution of thc aggrcgatc social product of capitalist countries, and the lion's share of their direct foreign investments and half of their home and foreign trade. The Soviet economist T. Ya. Belous, basing herself on these facts, has remarked:

The international monopolies, which have developed as a result of the export and international interlocking of capital, are exerting an ever more decisive influence on the processes taking place in the economies and politics of modern capitalism. Their role in the postwar capitalist economy has increased steeply and continues to grow rapidly. No phenomenon of any significance in the economic and political affairs of capitalist countries, and in the whole system of their international relations, can now be fully evaluated and understood without analysing the activity of these monopolies.

Among the long-term factors determining the specific features of the development of the postwar world capitalist economy, as we have already remarked, a very material role is played by the scientific and technical revolution developing in the postwar period. Monopoly capital is endeavouring to put its main stake on it in the struggle to consolidate its class positions and to raise the effectiveness of the development rates of production. This revolution, by encouraging growth and internationalisation of capitalism's productive forces, has not simply led at the same time to reproduction on an ever greater scale and to an unprecedented sharpening of all former antagonisms immanent in the capitalist mode of production but has also given rise to new contradictions that are unresolvable for capitalism.

The new phenomena include (1) the ever-growing contradiction between the immense growth possibilities opened up 66 by the scientific and technical revolution and the obstacles that modern capitalism is putting in the way of their employment in the interests of society as a whole; (2) the bundle of contradictions between the social character of production, rapidly increasing under the impact of this revolution, on the one hand, and the state-monopoly forms of controlling technical progress, on the other hand; (3) the rapid deepening (along with further growth of the antagonism between labour and capital in the countries of imperialism) of the gulf between the interests of the overwhelming majority of the nation and those of the financial oligarchy. The unnatural position, in which production complexes often serving more than one country remain the private property of a handful of millionaires and billionaires, is particularly obvious in tin's situation. The need for the replacement of capitalist relations of production by socialist ones is becoming more and more pressing.

The scientific and technical revolution is also furthering the rise of new forms of the manifesting of the unevenness of world capitalism's growth. There has been a marked deepening of the gulf between various countries and groupings in the sphere of scientific research, and in the area of the application of the latest, most effective results of technical progress in the national economy.

The U.S. monopolies broke away in this respect after World War II and forged ahead of the leading groups of the other capitalist powers. The 'technical gap' between the major and minor powers grew visibly, and also that between the imperialist countries as a whole and the developing countries.

By intensifying the spasmodic character of the economic growth of individual countries, the revolution at the same time inordinately heightened the imbalance of the various spheres of the capitalist economy. The contradictory character of the development of industry and agriculture, of the extraction and manufacturing industries, of the '65 traditional' industries and new ones, is growing.

Under imperialism, when scientific and technical advances may be introduced mainly in the most highly monopolised enterprises, considerable changes arc taking place m the industrial structure of postwar capitalism. A number of the succeeding chapters will be devoted to analysing all these problems.

CAPITALIST MODE OF PRODUCTION IN DEVELOPMENT COUNTRIES

The continuous interaction of the development of production and the world market had already come out in the early stages of the rise of the capitalist social formation, and the determinant place in this interaction, moreover, was taken by capitalist industry. As a result of his thorough investigation of extensive historical data, Marx came to the conclusion that when in the 16th, and partially still in the 17th century the sudden expansion of commerce and emergence of a new world market overwhelmingly contributed to the fall of the old mode of production and the rise of capitalist production, this was accomplished conversely on the basis of the already existing capitalist mode of production. The world market itself forms the basis for this mode of production. On the other hand, the immanent necessity of this mode of production to produce on an ever-enlarged scale tends to extend the world market continually, so that it is not commerce in this case which revolutionises industry, but industry which constantly revolutionises commerce.

Marxist political economy consequently looks upon the rise and development of the world capitalist market as a constituent of a broader process, *i.e.* the gradual conversion of capitalism into a universal socio-economic system. This methodological approach has enabled it to disclose the unequal character of the division of labour between countries that is inherent in this system, and the exploiter nature of commodity exchange and of capitalist international economic relations in general on the world market, based as they are on private property in the instruments and means of production.

The struggle for foreign markets has always been an urgent economic need for capitalism, since a striving to extend its sphere of influence without limit is inherently characteristic of it, in contrast to previously dominant modes of production. Because of the operation of objective economic laws, capitalist enterprise inevitably outgrows the boundaries of the community, region, or country, the economic exclusiveness and isolation of countries being broken down by commodity circulation itself.

As a result, capitalist production becomes inconceivable without foreign trade and world economic relations at a certain stage of its development, which in turn reflects growth of the social division of labour on an international scale. Marx stressed that capitalist production rests on the *value* or the transformation of the labour embodied in the product into social labour. But (his is only [possible] on the basis of foreign trade and of the world market. This is at, once the precondition and the result of capitalist production.

He expressed the same idea even more succinctly in the formula 'capitalist production does not exist at all without foreign commerce'.

All that in no way signifies, of course, that any capitalist enterprise indispensably works for an external market and sends what it produces beyond the local or national economy. At the same time it constantly experiences the effect of the already existing system of international business relations and is drawn into their orbit, if not directly, then indirectly. The capitalist, even the one who is wholly oriented on the home market, is forced to compare his own costs of production with world prices as well as with the market prices in his country. An influence of the same order has to be allowed for when an all-round appraisal is being made of the position of labour power in capitalist countries, including the spheres where its value depends wholly or predominantly on domestic factors of production.

Capitalism's international economic interconnections arose later, historically, than its domestic ones, the latter being original and basic. Their level and trend depended to a decisive degree on how far each nation developed its own 21 productive forces, division of labour, and internal dealings. That, however, does not eliminate the undoubted fact that extension of both the home and foreign market under capitalism represents different aspects of a single process; the boundaries between them are very mobile and arbitrary. As the world market develops the problem of the limits of the home market of each country taken separately proves to lie very tightly bound up with that of extension of its external economic relations. Hence one of the essential principles of Marxian researcli into the development of capitalism in scope and depth is that, when analysing international relations of production (including the capitalist class's colonial policy), it is not important where the boundary between the home and foreign market is drawn; what is important is that capitalism cannot exist and develop without constantly expanding the sphere of its domination, without colonising new countries and drawing old non-capitalist countries into the whirlpool of world economy.

The rise of capitalism's international exploiter relations is inextricably linked with the forming of a system of world economy appropriate to them.

The watershed on this point between Marxian and capitalist science is clear and distinct. Although the creation of this system was completed on the whole under imperialism, it by no means follows that its present-day history can be separated from the preceding history of the development of the capitalist mode of production and that imperialist exploitation of some countries by others begins only with capitalism's transition from free enterprise to the domination of monopolies. Many of the features of the international division of labour typical of the monopoly stage had already arisen and begun to take on a stable, long-term character during the transition of capitalist society to the machine stage of production. They were not brought about by chance, transitional factors but wore caused by deep-seated changes in both the home and external conditions of capitalist reproduction.

The analysis of international economic inter-relationships made in their day by the founders of Marxist political economy indicated that it was large-scale machine industry that prepared the ground for the subsequent formation of a world economy. Machine production, as Marx and Engels pithily put it, produced world history for the first time, insofar as it made all civilised nations and every individual member of them dependent for the satisfaction of their wants on the whole world, thus destroying the former natural exclusiveness of separate nations.

It is through the international division of labour on the world market that the tendency to bring countries together economically, constitutionally inherent in capitalism, is manifested. The kernel of this pattern, however, discovered by Marx, is that this tendency, while progressive in itself, led at the same time, because of capitalist methods and the forms of economic association of the different countries, to consolidation of the division of the world into industrially developed countries, and economically backward agrarian and primary commodity producing countries exploited by them. The antagonism between these two main groups of countries, which are opposite poles of the world capitalist economy, has become one of its main contradictions.

This contradiction has become specially acute in the stage of imperialism, and a very important factor in subsequent origin and development of the crisis of the colonial structure of the capitalist economy, a crisis that is irreversible for the monopoly capitalist class. The break-up of the colonial system after World War II, and the marked deepening of the internal contradictions of world capitalism, which are undermining the historically formed foundations of its system of unequal international economic relations, have put on today's agenda, with all urgency, the issue of eliminating the exploiter system of division of labour that completely dominated the world economy in the past.

It would be wrong to treat this task (which corresponds to the vital interests of all nations without exception) as the need to liquidate the progress already made in the internationalisation of social production. The anti-imperialist struggle for genuine national independence, just like the transition of countries to a higher socio-economic system, does not eliminate the objective need to develop the social (including the international) division of labour, and leads to a change in and perfecting of the forms in which this essentially progressive process is manifested. It is self-evident, Marx wrote to Ludvig Kugelman on 11 July 1868, that this necessity of the distribution of social labour in definite proportions cannot possibly be done away with by a *particular jiirin* of social production but can only change the mode of *its appearance.*

The underlying principles of a Marxist-Leninist analysis of the ways and causes of capital's internationalisation of social production have not lost their theoretical value over the years. The topicality of many of them, moreover, is growing, especially in today's situation of the coexistence and opposition of two world social systems and of the natural extension of economic and technical co-operation on the basis of mutual advantage and equal international division of labour.

In this situation capitalist ideologists are trying in every way to prove the 'erroneousness', in particular, or at least 'obsoleteness' of the Marxist-Leninist political economy's analysis. Above all they endeavour to refute its conclusions about the irreversibility of tendencies towards a steady deepening of capitalism's internal contradictions. But the whole long history of capitalist economics, and of its mistakes, its constant wavering between one conception and another, its fallacies and failures, shows that theories that consider the international market relations of capitalism to be legitimate and that are abstracted from a socio-economic analysis of the production sphere, at best contain superficial analogues of the real trends of development. In the main they bring out the consequences but not the causal connections of this development.

At the same time it would clearly he an oversimplification of things to consider that the theoretical findings of capitalist political economy in the sphere of the world economy's market relations are in general isolated from real analysis of the reproduction processes. In a numher of cases they more or less adequately reflect important features of the backlash of these relations in the production sphere. And this effect determines many of the essential features of the capitalist economy at the various stages of its development. In the last instance (Engols wrote to Conrad Schmidt on 27 October 1890) production is the decisive factor. But as soon as trade in products becomes independent of production proper, it has a movement of its own, which, although by and large governed by that of production, nevertheless in particulars and within this general dependence again follows laws of its own inherent in the nature of this new' factor; this movement has phases of its own and in its turn reacts on the movement of production.

All this calls for close attention to research both into the ways and determinant tendencies in the genesis of the world market and into its place in the process of extended capitalist reproduction. Not only is such research of educational value historically; it is also very necessary for an analysis of current changes in the structure of the world capitalist economy. A systems analysis of its inner and external interconnections in today's international situation cannot be made sufficiently full without detailed allowance for tiie general patterns and long-term trends of development of these interconnections.

Lenin more than once stressed the need to study broad socio-economic phenomena in historical retrospect, since the most important thing if one is to approach this question scientifically is not to forget the underlying historical connection, to examine every question from the standpoint of how the given phenomenon arose in history and what were the principal stages in its development, and, from the standpoint of its development to examine what it has become today.

There were relatively broad trade and financial relations between separate countries and peoples, of course, in the socio-economic formations that preceded capitalism, but it was capitalism that first formed a world market which played a progressive, revolutionary role in the subsequent increase of mankind's productive forces and growth of its social production. Marx, attributing paramount importance to that, wrote that 'the specific task of bourgeois society is the establishment of a world market'. It was with the formation of a world market that the level of development of capitalism began to be marked by the degree of growth both of domestic and of international market relations.

There was a qualitatively new leap in the shaping of these relations associated with the vast increase in trade and colonial expansion of the young European capitalist class. This expansion became a decisive factor in the primitive accumulation of capital. The world market thus created important material preconditions for victory of the capitalist modeof production over the feudal.

At that time the merchant capital of European powers undoubtedly prevailed in international economic relations, but its dominance on the world market began to decline as the capitalist mode of production was consolidated in the countries that were the basis of its existence. The subordinating of merchant capital to industrial, natural to the new mode of production, was also manifested then in the realm of international trade. As a result the material conditions were laid for an ever increasing intensification of the action of the mechanism of the economic exploitation of all the countries of the world by a handful of 'advanced' countries. And the wider the gap in levels of labour productivity becomes between this handful of advanced countries and the economically backward ones, the greater is the exploitation suffered by the latter on the world market. This pattern was expressed with maximum clarity in the following formula of Marx: The favoured country recovers more labour in exchange for less labour, although this difference, this excess is pocketed, as in any exchange between labour and capital, by a certain class.

In that way, in particular, the main antagonistic contradiction of the capitalist mode of production, *i.e.* that between its constantly increasing social character and the private capitalist forms of appropriating the results of social labour, thus found expression within the world market.

The putting of capitalism onto the rails of large-scale machine industry during the industrial revolution, which was completed in the second half of the nineteenth century, meant the beginning of a new, higher stage in the development of world economic relations. Large-scale industry in fact shaped the world market of the epoch of mature capitalism. And its dependence on external economic links and the international division of labour in turn also grew immensely. As Karl Marx said in his day: Large-scale industry, detached from the national soil, depends entirely on the world market, on international exchange, on an international division of labour.

SITUATION THE ANTAGONISTIC CONTRADICTIONS OF THE WORLD MARKET STEMMING

In this situation the antagonistic contradictions of the world market stemming from the basic economic contradiction of capitalist society, have become an ever more serious obstacle to a further rise of social production. One of the most convincing bits of evidence of that were the world economic crises that were converted into a constitutionally inalienable feature of the whole economic system of capitalism. Years of comparatively rapid industrial boom began to alternate with regular consistency with periods of stagnation and direct destruction of already created productive forces. This cyclicity, determined by the objective laws of capitalist production, was intensified in turn by industry's mounting dependence on the foreign market. It was their comprehensive study both of the internal, home factors of the cyclic development of capitalist business and of the international ones that enabled the authors of classical Marxist political economy to draw the conclusion, confirmed by the whole subsequent course of social development, that economic crises would remain, while capitalism existed, one of the incurable ills determining its historically transient character.

For more than a century and a half now, since 1825, capitalism has periodically suffered from cyclic crises. All the attempts of capitalist economists to work up prescriptions to cure this illness without eliminating its basis, namely its exploiter relations of production, have unfailingly proved unsound and bankrupt. The capitalists as a class have demonstrated more and more clearly with the course of time their incapacity to control the productive forces of society rationally not only within the context of the separate national economies but also and especially on the scale of the world economy. The world cyclic crisis of 1974–75, the greatest since the war, which developed under an unprecedented sharpening of several of modern capitalism's most important world problems, *viz.* the monetary, energy, raw material, and food crises, was convincing evidence of that indisputable fact in our day. It swept almost all the main centres of capitalism simultaneously and in turn interrupted their economic development. During the crisis the monopoly capitalists as usual exerted immense efforts to shift its most serious consequences onto the working masses of their own countries and the peoples of the agrarian and primary producer periphery.

As capital internationalises society's productive powers the connection between the reproductive processes of the individual countries (including those that are the epicentres of world crises under the impact of the economy's cyclic movement) is becoming closer and closer. Therefore, when we are determining the objective patterns of capitalism's development, the question of the interaction of the national cycles within the context of a single world cycle (study of which was begun by Karl Marx) is posed in all its acuteness. We shall dwell in detail on the features of the cyclic development of today's world capitalist economy later, here it is necessary to stress the following from the standpoint of the matter we are examining.

Despite the fact that the cyclic synchronousness of the movement of capitalist production and of the world market already lias a history of 150 years, a line of refusing to recognise the general patterns of this movement still clearly dominates capitalist economics. When, however, the course of the world cycle is studied, its development is mainly considered as a kind of arithmetical sum total of the results of cyclic fluctuations within the national economies taken separately.

As a result it comes about that the specilic features of national development are, as it were, the sole decisive factor in the shaping of each world cycle. In that case the role of the system of international socio-economic connections and of the internationalisation of social production, which in essence convert these countries into links in a single economic chain, is patently underestimated.

Since the beginning of the 70s, it is true, because of the steady exacerbation of modern capitalism's international economic problems, an increasingly negative attitude towards sucli a 'narrowly nationalist' approach to analysis of the main world crisis processes (including its cyclic development) has been traceable in capitalist economic literature. Sometimes, while justly noting the close connection between these processes, and in particular between national overproduction crises, within the context of the capitalist economy, most capitalist economists as usual, however, prefer to identify their prime causes with the processes taking place at any definite moment in the sphere of the international market and political relations. They thus willy-nilly leave aside objective study of the system of international exploiter relations of production that underlies the deep internal contradictions of the world capitalist economy, which are inexorably undermining its main foundations.

The political dependence of most of the countries of the world on the main centres of capitalism has played an extremely important z'ole in the long history of the building up of this system. The structure of its unequal, exploiter relations was also moulded by that dependence in the imperialist stage, the crisis of which developed with particular force in the second half of this century as a consequence of most colonial countries' winning of national independence.

DEVELOPMENT OF AGRO-INDUSTRIES IN INDIA

'Agro-industry' is an omnibus expression. It could cover a variety of industrial, manufacturing and processing activities based on agricultural raw materials as also activities and services that go as inputs to agriculture. The agro-industries corporations, set up during the 'sixties in most states, have mainly been engaged in supply of farm machinery, fertilizers, seeds and other modern inputs available to farmers. Processing of agricultural produce is, however, a well-known agro-industrial activity. Besides the two-way linkages to agriculture, one would need additional criteria to classify agro-industries. To make no distinction between the nature of economic activity involved in spinning and weaving in modern mills and the traditional village weaver working with home-spun yarn would, for obvious reasons, be not justified. Similarly, while agriculture is undoubtedly the main source of raw materials for cotton textiles, it would appear ludicrous to classify Ahmedabad and Bombay as the centres of agro-industries, because both host cotton textile industry.

The same would be true of Calcutta, known for its jute mills, and Hyderabad with its large tobacco industry. From the viewpoint of backward linkages to agriculture also, it would appear odd to label fertilizer and tractor manufacturing units as agro-industries. The need for a more rigorous definition of agro-industries is obvious enough. While doing so, it may be necessary to take note of the nature and extent of two way linkages, location, capital, technology, entrepreneurial characteristics and features of the market structures. The attempt to outline the definitional contours of the concept of agro-industries also calls for spelling out the purpose and the context in which the classification is being evolved.

Traditionally, Indian agriculture drew most of its inputs locally from the village and the farm. Modern inputs, like chemical fertilizers, pesticides, tractors and other agricultural implements, pump sets, diesel, gas and power, had little import in the overall agricultural inputs. Agro-industries were essentially perceived as first level post-harvest processing of farm produce. Agro, village, cottage and rural industries meant the same set of economic activities. The expressions were interchangeable. Even when agro-industries have been assigned a special place in the successive Five Year Plans, a good deal of confusion continues to persist regarding their coverage. It is not very clear whether agro-industries are to denote only the activities directly related to agriculture or the total farm output and related activities. A broad-based classification on this would help appropriate classification of dairy farming, poultry, piggery and other farm activities. Similarly, a clear-cut view should be taken on whether tea, coffee, rubber, spices and other plantations are to be classified as a category separate from agricultural activity.

In a typical Indian village, economic activities were and, to a considerable extent, even today are generally associated with a particular 'caste' and community, which has practised a particular profession for generations. The nature of activity undertaken was, in general, hereditary and occupational mobility within the village nearly non-existent. Generally, each village had at least one household each of blacksmiths, carpenters, weavers, potters and those engaged in oilseed crushing. Agricultural and non-agricultural activities had direct and strong organic links within the village. Landowners needed manpower to cultivate their lands and help them in the household. They also needed professionals to take care of their plough and other agricultural implements; in return, the village artisans were paid in the form of grains and other farm produce each season. The relationships, sanctified by socio-economic traditions, were thus institutionalized. A typical Indian village in this sense was an organic whole, in which, while each activity had its own distinct identity, there was a considerable degree of inter-dependence.

Under the inspiration of Mahatma Gandhi, the national struggle for India's political independence witnessed a concomitant struggle for the preservation, protection and encouragement of rural industries. The unequal competition from cheap mill-made products threatened employment and livelihood of the rural artisans and craftsmen. The preference for *Khadi vis-a-vis* mill cloth, and cottage and village industry products *vis-a-vis* the urban and mill-made products was motivated from a realization that the experience of urban handicraft centres might also get extended to rural non-agricultural activity and the cottage and village industries. Gandhi put a premium on simplicity in life style and consumption. The Gandhian strategy for Indian development was linked to enhanced utilization of the vast mass of surplus manpower and its active involvement in production processes. The Gandhian ideology was not only economic but also social and political. In the Gandhian idiom, cottage and village industries represent a support structure to a life style that is more moral than economic. Villages keep the workers in close touch with balmy open spaces and nature in all its peace and pleasantness.

The view, that village industries and crafts are an important part of rural life and should be vigorously protected to ensure sustenance to a self- reliant village, is essentially an outcome of the traditionalist philosophy, articulated so forcefully by Gandhi. It was, infact, a reactive approach — a defence mechanism against the onslaught of the British industry. Completing the first stage of industrialization, the British industry was then looking for new markets in the British colonies for its products along with linkages for obtaining raw materials. India was a good opportunity on both counts. The task became easier with improvements in roads, introduction of railways and use of rivers as navigational channels. India's population was large and it had a rich tradition of crafts and village industries. A variety of industrial products were already in use. Low priced substitutes could find ready markets here. On the other hand, India's agricultural base was vast and land was not fully exploited. With a little effort, India could be developed as a dependable source of raw materials. The British imports could be contained, if India had gone in for establishment of its own processing industries and other manufacturing capabilities. That would have made it difficult for the British industries to gain hold on the Indian market. This, however, was by no means possible in a colonial set up. The discrimination against the Indian entrepreneurs is well documented. Imported mill products were cheap as also of uniform and better quality. In spite of the fact that raw materials for the British mills were obtained from India, the low processing costs made it possible for the British industry to capture the Indian consumer market. This was particularly so in the case of cotton textiles. The British industry expanded at the cost of the Indian small scale and village industries. A direct consequence of the loss of market was unemployment of the non-agriculturalists.

Indian political leadership identified itself with the Indian artisans and craftsmen, who were being adversely affected by the competition from abroad. The call for *Swadeshi* and boycott of foreign made goods can be well understood only in this context. *Swadeshi* was the strategy to protect employment for millions of the Indian non-agricultural workers. Adoption of traditional consumer goods, made by the cottage and village industries, got associated with the Indian nationalist sentiment and the national struggle for political independence. The lead political party, the Indian National Congress, in fact, made it obligatory on all 'active' party workers to wear hand spun and hand woven cloth. To wear *Khadi*, use *ghani* oil, prefer hand-pounded rice, and buy hand made and locally produced shoes became a symbol of patriotism, nationalism and political commitment against foreign (British) industries. It was thus a preference for hand-made consumer items.

Swadeshi was a positive and powerful assertion against foreign and mill-made products. Gandhi had the insight to understand well that the *Swadeshi* movement could sabotage the economic interests of the British as also help mobilize mass support for the national struggle for independence. The movement against mill-made products would not have held well, if the consumer goods industry in the mill sector of India had already been well developed. It would have been difficult to pursue this line, if Indian mills and textile industry were owned and managed by Indian economic interests. This was, however, not the case, for most of the large industrial complexes were under the British Managing Agents.

The support to rural industries and hand-made products was not only a reaction and a defence but tended, in practice, also to become a basis, and a near habit for a good many nationalists to oppose modern industrialization. Associated inevitably with this philosophy was a choice in favour of the old but less efficient techniques of production. The self-reliant village philosophy might have had an element of romance but it would be wholly out of place in the present context, when space is drastically and distances, in terms of time, are getting drastically reduced because of improvements in transport and communication technologies. Each village seeking to develop self-sufficiency in as many products as possible, implied non-aggregation of demand at the state, regional or national levels. The fragmented demand would not permit adoption of production processes that thrive on the economies of scale and permit specialization. Most of the traditional techniques of production are labour intensive; hence employment-oriented. Continuance of emphasis on the traditional way of life, technology and the value system does imply acceptance of monotony, drudgery at the work place and low productivity.

The vision of agro-industries/rural industries, as projected by the political leadership during the pre-independence era, had some serious limitations, the most significant ones being that it did not project itself to the likely impact of: (i) the spread of literacy and technical education, (ii) availability of alternative technologies, (iii) the growth of mass media, (iv) the changing aspirations of the people and the youth in particular, and (v) easy availability of power and electricity. The truth is that agro and village industries were seen in terms of production in a traditional village and not the village of the future in a free and modern India. It was not comprehended that occupational structure, employment of women and a variety of gainful employment opportunities, especially in the service sectors, could grow rapidly with cheap and efficient transport, recognition of environmental and other factors associated with large industrial complexes, and urbanization. Agro-industries, as traditionally understood, could also not accommodate regional specialization to exploit comparative and locational advantages. It also remained un-appreciated that techniques of production are not always independent of the socio-economic system. Continuance with the traditional technologies could only help provide protection to the caste system and whatever goes with its justification.

Agro-industries have also been viewed as a safety valve that needs to be built within rural areas to absorb surplus labour and provide relief to the problem of large scale disguised unemployment. A good many Indian official reports and other important writings make a plea for agro-industries in the context of rural-urban migration. Absence of employment opportunities within the village, it is suggested, is the main push factor responsible for the rapid movement of youth towards cities. Emergence of slums in metropolitan towns of the country and arrival of unattached young without gainful employment is the direct and inevitable consequence.

These developments have a variety of social, law and order, and political implications. During the 'fifties, Indian planners entertained a hope that promotion of agro-industries would help avoid furtherance of industrial concentration and achieve a more balanced regional dispersal of industrial activities and employment. They also hoped that the widely distributed industrial activity could help reduce pressure on transport and other economic and social infrastructure.

The traditional viewpoint did not, however, realize that migration of population is not a mere change of residence and profession. It is also a process that permits inter-mix of people with differing socio-economic and cultural backgrounds. Industrial centres become the crucibles and melting pots where differing cultures merge and fuse with one another to give birth to new cultures and new personalities with more rational, logical and scientific value systems than those prevailing in the feudal and caste ridden traditional village societies. Value systems and social institutions change with urbanization. It is no surprise, therefore, that urban centres always attract the poor and members of the lower castes. Urbanization in India has promoted vertical occupational mobility. An untouchable could become a cook in a city environment. Could he be ever permitted to become a cook or socially accepted as such in a traditional village? To the extent agro-industries were being suggested to preserve the identity of the village, its social structure and the production system, it was a denial of the merits that would come with urbanization. In India, urbanization has rarely, if ever, been appreciated; if at all, it has been accepted as a necessary evil.

Soon after India's independence the Congress Party constituted the Economic Programmes Committee to provide a broad direction to the Congress Governments at the Centre and State levels. The Committee, headed by Jawaharlal Nehru, reported in January 1948. In its recommendations on industries it observed:

Industries producing articles of food and clothing and other consumer goods should constitute the decentralized sector of Indian economy and should, as far as possible, be developed and run on a cooperative basis. Such industries should for most of the part be run on cottage and small scale basis.

This was a large area earmarked for rural, cooperative and small scale industries. The general direction indicated for state intervention was for imposing restrictions on large scale manufacturing of most consumer goods while extending support to traditional systems of production. The Indian

National Congress took over the reins of power after India's independence in 1947. Most of the Congress leaders came to occupy responsible positions in the Central and State Governments. It was, therefore, expected that the sanctity bestowed on *Khadi* and village industries by Gandhi would preserved and adopted unhesitatingly as a national and public policy in free India. It was also expected that the ideals of building self-reliant villages was a commitment that had to be discharged faithfully by the Party and the followers of Gandhi. Because of the then prevailing political environment, especially the place and the role Gandhi played in the struggle for independent, there was no serious debate on or critical scrutiny of the issues Gandhi was associated with and had expressed his clear views. There were, however, many who were extremely critical of the Gandhian philosophy for socio-economic development of the country.

Since Gandhi stood for a decentralized administrative system, in which village was to be the basic unit of management, planning and administration, it was proposed by the Gandhians that it should be made obligatory for the state to pursue Gandhian ideals with regard to village and cottage industries. Reacting to them while speaking on the Directive Principles of State Policy, Dr Bhim Rao Ambedkar observed that Indian village to him was "a sink of localism" and "a den of narrow minded-ness". Similarly, opposition from many other veterans that did not permit adoption of another Directive Principle requiring the state to endeavour "to promote cottage industries on cooperative lines in rural areas".

The modified version reads: "the state shall endeavour to organize agriculture and animal husbandry on modern and scientific lines". The reference to "cottage" and "cooperatives" was quietly dropped.

The post-independence years represent a period of turmoil, when the slogans and emotional ideals of the period of India's struggle for independence came under critical scrutiny. It was natural. If none had questioned Gandhi and his philosophy earlier, many now doubted the relevance of Gandhian concepts in independent India. As long as Indian industry was dominated by the British, it made political sense to damn machine; but in the changed situation, machines needed as much a priority as any other programme of development. The organized industry was no more controlled by the British nor was there a threat to the Indian village industries from the textile mills located in Britain. During the post-War period a number of foreign companies changed hands. The character of conflict had changed. Instead of the foreign and organized interests *versus* Indian artisan, it was now the large Indian industrialists *versus* those who supported village and small scale industries. Both sides were represented by Indians.

The pre-independence view that the cottage and small industry faced direct competition from the big industries stood revised. The big was not a substitute for the small and *vice versa*. Both were to grow; in fact, in a mutually supportive manner. With a view to reducing the areas of direct confrontation, the

Government of India adopted a number of specific measures such as a favourable treatment to *Khadi*, cottage and village industries. During the period, 1952 to 1954, the All India Khadi and Village Industries Board and a Board each to promote silk, coir, and handicraft, handloom and small scale industries were instituted. These Boards were required to recommend general policies and prepare action plans for promoting activities in their respective areas through preference in Government purchase and distribution of raw materials, fiscal and monetary concessions, and supportive administrative policies. There was, however, no special category of industries called agro-industries. The earlier position thus stood revised. Machine was no more a synonym of colonial exploitation. Indications of the change in emphasis surfaced up soon. *One*, instead of taking the responsibility of promoting and protecting village and cottage industries directly, the Central Government shifted it to State Governments. *Secondly*, a somewhat ambiguous position was taken in statements such as "cottage industry could produce industrial components to be assembled at the factory level". The role of the small and large units was thus envisioned not as one of conflict but of complementarity. The first official declaration of Independent India observed:

The healthy expansion of cottage and small scale industries depends upon a number of factors like the provision of raw materials, cheap power, technical advice, organized marketing of their produce, and where necessary, safeguards against intensive competition by large scale manufacture, as well as on the education of the worker in use of the best available technique. Most of these fall in the Provincial sphere and are receiving the attention of the governments of the Provinces and the States.

It added that the Central Government would investigate how far and in what manner these industries could be coordinated and integrated with large scale industries. The Industrial Policy Statement (1948) promised to examine:

... how far the textile mill industry can be made complementary to rather than competitive with the handloom industry, which is the country's largest and the best organized cottage industry. In certain other lines of production, like agricultural implements, textile accessories, and parts of machine tools, it should be possible to produce components on a cottage industry scale and assemble these into their final product at a factory.

The First Five Year Plan made a distinction between village industries, small industries and crafts. Village industries were defined in terms of activities which are, in the main, an *integral part of the village economy*. The small industries and crafts were distinguished on the basis of (i) *traditional skills* and crafts, and (ii) the ones which have recent origin and have an intimate *connection with the corresponding large scale industries*. It was recognized that while the then existing village industries were of a rudimentary character, rural electrification was likely to transform them significantly. Agricultural development and village industries were looked upon as Siamese twins that could hardly be separated from each other. The First Plan also visualized that an increase in agricultural production would raise farmers' incomes and expand opportunities for processing raw materials in the villages. In order to satisfy the increased demand from the farmers, more persons could get employment within the village itself. The First Plan envisaged as below:

Amenities in rural life such as supply of pure drinking water, street lighting, sanitation, hospitals, recreation grounds, community centres and roads increase the field for village industries. The possibility of turning waste into wealth, for instance, production of gas from cow dung and other refuse of the village through gas plants in so far the operations prove economic, production of bone manure through bone digesters, soap making out of non-edible oils, etc will further provide scope for the development of village industries.

The Planning Commission decided to draw up Village Industry Programmes in consultation with experts. For this, the industries covered were:

- village oil industry;
- soap making with neem oil;
- paddy husking;
- palm gur industry;
- gur and khandsari;
- leather industry;
- woollen blankets;
- high grade hand-made paper;
- bee keeping; and
- cottage match industry.

Other industries assigned special priority were: *Khadi*, coir, sericulture, fisheries, forests, dairying and horticulture.

It is not necessary in this paper, given its limited focus, to dilate upon the Second Plan strategy. In addition to stressing the role of heavy industry, the Second Plan also assigned a special place to rural, cottage and small industries. It envisaged that the expanding demand for consumer goods sector would be met from outside the large units. This would reduce pressure on the capital and the limited savings of the economy and the strategy would fit in well with the need to expand employment opportunities. Explaining the Second Plan strategy, Mahalanobis observed:

In view of the meagreness of capital resources there is no possibility in the short run for creating much employment through the factory sector.... Now consider the household and cottage industries. They require very little capital.

The objectives of the Second Plan programmes and the Industrial Policy Resolution, 1956, were to create :

... immediate and permanent employment on a large scale at a relatively small capital cost, meet a substantial part of the increased demand for consumer goods and simple producers' goods, facilitate mobilization of resources of capital and skill, which might otherwise remain inadequately utilized and bring about integration of the development of these industries with the rural economy, on the one hand, and large scale industry, on the other. They also offer a method of ensuring more equitable distribution of the national income and avoiding some of the problems that un-planned urbanization tends to create. With improvements in techniques and organization, these industries offer possibilities of growing into an efficient and progressive decentralized sector of the economy, providing opportunities of work and income all over the country.

It is obvious that the cottage and small scale industries were visualized as a panacea for most economic ills. It was probably the emotional appeal and romance of the heroic memories of the national struggle for independence that mesmerized the planners so that very few questioned the *raison d'etre* of small scale and village industries in the long-term scenario of national development.

There were many Gandhian thinkers, who were unhappy at the manner in which official policies and plan programmes betrayed, according to them, the confidence of the poor. This is best illustrated by a Planning Commission note of August 1961, which had been prepared by Jaya Prakash Narayan and outlined the Gandhian philosophy of rural industrialization. The note underlined:

- Plan efforts have contributed very little to (a) create more employment, and (b) enhance wealth of the rural communities and raising the standard of living, particularly of the backward sections, who constitute the overwhelming majority of India;
- Given the high population growth and limits to absorption of labour in agriculture, rural industrialization alone can provide an effective remedy for the rural unemployed;
- Industrialization need not be only an urban phenomenon. It can equally be a rural one. Its nature, however, would be different.
- Rural industrialization should not be confused with setting up of a few large industries in rural areas. Rural industrialization must mean an even spread of industries throughout the countryside all over the country;
- Nor should rural industrialization be limited to mean what at present are termed as 'rural industries' or to only agricultural industries, such as the processing of agricultural commodities. There can be and should be infinite variety of industries for rural India;
- Rural industrialization will have to be based on two factors: (a) local resources, both human and material, and (b) local needs. Local does not, however, mean only a village; it might mean a village, a group of villages, a block, or a district;
- There are to be no pre-conceived limitations or inhibitions of a doctrinaire or sentimental type in this regard to such matters as the

use of power and technology. The only concern should be: (a) to maintain a balance between employment and efficiency, and (b) to preserve and nurture certain social values as those of political and economic self-government, minimizing, if not eradicating, economic exploitation; promoting economic and social equality, and preservation and development of producers' personality. Jaya Prakash suggested that the aim and total long term effect of rural industrialization should be to convert the present lopsided purely agricultural communities into balanced agro-industrial communities.

The Planning Commission identified consequently 40 rural areas for intensive development of small industries. The primary objectives of its programme were to:

(a) Bring about a cooperative agro-industrial economy; and create employment opportunities to enable a higher standard of living; and
(b) Mobilize rural communities and seek diversification of rural economy in a manner that contributes to the welfare of the landless and the weaker sections of village communities.

Rural industrialization was then seen to have two components, namely (i) location, and (ii) linkages with large industries as ancillaries. The Rural Industries Programmes were to cover all kinds of small industries and processing industries based on agriculture. It was recognized that:

With the increase in the production of cereals, pulses and a number of cash crops like sugarcane and oilseeds visualized in the Third Plan, there will be considerable scope for the expansion of processing industries in rural areas. With a view to providing fuller employment and strengthening and diversifying the rural economy, it will be desirable to develop these industries to the maximum extent in the decentralized and small scale sector and on a cooperative basis.

Different varieties of the decentralized sector (cottage, rural, small or agro) continued to enjoy a special place in the successive Five Year Plans. The inherent strength and weakness of the policies towards small and village industries are now better appreciated. The cottage industries and products of the rural crafts have found a good market among the Indian urban elite; a large part of the consumer goods market has, however, been captured by the organized and large enterprises. The official patronage to small scale, rural or cottage industries does not appear to have made a noticeable impact on the problem of rural-urban migration. The establishment of cottage and village craft emporia and training of artisans and design workshops has also not mitigated the situation. The *Khadi* and handloom sectors do seem to continue providing an additional source of income in villages. It is, however, doubtful if these industries, left to themselves, have the inherent strength to face competition by the modern mill sector. The problems responsible for the poor progress of village industries were well summed up by the Third Five Year Plan (1961-62 to 1965-66):

Rural artisans are usually dispersed in a large number of scattered villages and this, combined with their low standard of literacy and poor economic condition, is a considerable impediment to rapid implementation of development programmes. Among the other factors responsible for the slow progress of village industries' programmes have been the general lack of previous experience in regard to the development of these industries, lack of trained and qualified staff, location of production centres in unsuitable places, lack of adequate funds and organization for procurement of raw materials in bulk and failure to introduce more efficient techniques of production. Even such technical improvements as were introduced did not go far enough to secure a material increase in productivity. They did not, therefore, gain general acceptance.

Two schools of thought seem to have dominated the policymaking levels during the early 'sixties, — one represented by the official circles and the other by those who sought to have integrated rural development. The main concern of the official policy has been to pursue Plan objectives of growth and sectoral targets, whereas the non-official view is replete with a deep concern for the rural development, in general, and the rural poor, in particular.

The 'sixties witnessed the beginning of the green revolution in some parts of India. In the Punjab, Haryana and western Uttar Pradesh, agricultural output per hectare rose markedly due to the enhanced canal and well irrigation, widespread adoption of new and improved seed varieties, enlarged inputs of chemical fertilizers and use of pesticides. While managerial practices are important, it is an undisputed fact that the green revolution was a direct consequence of high levels of agro inputs per unit of land.

The enlarged inputs were not obtained from the farm itself or from traditional sources. The switchover to electricity, diesel and pumpsets was almost dramatic; the high-yielding seeds were brought in from research centres; and tractors and agricultural implements, supplied by national and international sources. The green revolution brought Indian agriculture in close contact with industry, the nature of agro-industry relationships extending themselves to supply of industrial inputs instead of agriculture playing the raw material supply function only. The prosperity of farmers was also bound to generate new consumer demands produced by industry. The demand for a variety of industrial inputs had to be satisfied, if agricultural development was to be optimized. With a view to reducing problems of procurement of industrial inputs for agriculture, the State Governments were advised by the Centre to set up Agro-industrial Corporations. With the following principal objectives:

(a) Promotion and execution of industries undertaking production, preservation and supply of food;
(b) Enabling persons engaged in agricultural and allied pursuits to own the means of modernizing their operations;

(c) Distribution of agricultural machinery and implements as well as equipment pertaining to processing, dairy, poultry, fishery and the industries connected with agriculture;

(d) Undertaking and assisting in the distribution of inputs for agriculture; and

(e) Providing technical guidance to farmers and persons concerned with agro-industries with a view to enabling efficient conduct of their enterprises.

The Agro-industrial Corporations were to promote agro-processing and generate additional employment opportunities in rural India. These objectives did not, however, find a priority in their actual working, for they chose the easier course of promoting sales of tractors, agricultural machinery, fertilizers, pesticides, etc. None could have been happier than the large enterprises manufacturing tractors and other agricultural machinery, when they found these state-sponsored corporations were most willing, even anxious, to undertake marketing for the private sector. Most State Agro-industries Corporations make profits. In some states, beginning with the early 'seventies, subsidiary corporations were also set up to provide cold storage and crop processing facilities. The variety of activities undertaken by the Agro-Industries notwithstanding, their the main operations continue to be organizing modern inputs to agriculture.

The official efforts at promoting village, rural and agro-industries were grossly inadequate compared to the magnitude of the task involved. They were only too thinly spread both in relative and absolute terms. Except for the Second Plan, the expenditure on

Khadi and village industries, including the small scale industries, has never exceeded three per cent of the total expenditure by the Centre, State and Union Territories.

It was four per cent during the Second Plan. Actual expenditure during the Seventh Plan and the outlay in the Eighth Plan are less than one-and-a-half per cent. Understandably, the share of the *Khadi* and village industries sector has been less than one per cent during most of the period since independence.

In relative terms, the allocation of funds for Khadi and village industries, as per cent to total Plan outlay, came down from 0.8 in the First Plan to 0.3 in the Seventh Plan even though it had gone up to 1.8 in the Second Plan. This is a cause for concern, especially when the Government swears by priority to rural development and creation of employment opportunities. The actual release of funds for Khadi and village industries (KVI) has been slightly even less than the Plan allocation throughout except during the Sixth Plan period, when it was slightly more. The general problem in the country is not poverty *per se* but poverty born out of unemployment. The KVI sector is the cheapest option for generation of employment, as it requires only Rs. 5,000 for Khadi and around

Rs 10,000 for a village industry per employment. The low investments on the sector, spread thinly over different states, are accompanied by an absence of any attempts to arrive at a relationship with regional resource endowments.

The 'eighties witnessed a strong plea for promotion of agro-industries in India. The orientation and the context of the assertion, however, has been vastly different from the arguments of Gandhi, Karve, Mahalanobis and Jaya Prakash Narayan. Agro-industries of the 'eighties are essentially understood in terms of food processing industries. The arguments, briefly put, are:

One, in spite of a very low per capita income, India has an estimated population of around 80-100 million constituting the middle upper class that supports a reasonably high consumption standard. This offers a large market for modern durables and agro-based products, especially semi-processed and convenience foods.

Two, in addition to the large internal market there exists a huge unexploited potential in the international market, where India has competitive edge over many other supplier countries.

Three, growth of food processing industries would provide expanding demand for farm produce, vegetables, fruits and other greens that would help improve agricultural incomes.

Four, the industry would give consumers in having access to vegetables, fruits and other farm products throughout the year and, equally important at low and stable prices. This would, of course, mean better returns and incentives to Indian farmers. *Five*, establishment of modern plants with sophisticated technology would help reduce crop wastage due to seasonal gluts and the perishable nature of farm products.

Six, urban centres are witnessing a substantial change in the intensity of woman employment. In families with both husband and wife doing formal jobs, as per modern life-style, there is a growing potential for consumption of convenience and semi-processed foods.

And *seven*, the demand for processed food is likely to be enhanced because of the growing problem of obtaining full-time household assistance. The case for establishing food processing industries rests on the premise that there exists a large potential for products of the industry at home and abroad. It is a matter of more than coincidence that the initiation of the interest (in India) in promoting agro-industries has been simultaneous with the efforts made by some transnational corporations to seek entry into the Indian food and soft drink market. The post-1985 period has witnessed a substantial change in the official policy towards technology import and role of private foreign capital, in general, and in the food industries, in particular. The change is evident from the fact that, of the 111 foreign collaborations in food related industries approved during 1951-1985, more than 60 per cent of the agreements were for technical know-how and only 40 per cent for financial collaboration. The distribution of the collaborations in 35 years in term of the nature of processes shows that nearly 40 per cent of the collaborations were for "plant and machinery". As a single product area, "marine" was the main sector, which enjoyed a dominant position. Fruits, vegetables, edible oils and other food-related products did not have much of a place.

The 'eighties witnessed a keen interest in investments in the area of food processing and soft drinks. The TNCs visibility in this area is indeed a marked one. For instance, Pepsi entered into collaboration with Punjab Agro Industries Corporation and the Tatas to establish processing facilities for tomato juice and paste along with soft drink concentrates.

Though a failure, General Foods of US also entered India during this period. Kellogg has evinced interest in production of breakfast foods. Nestle, known for its interest in coffee, has started marketing "Maggie" convenience foods, ketchup, chocolates, etc; Hindustan Lever, the first entrant to the hydrogenated edible oil industry in India, handed over the *Dalda* production and marketing to its sister company, Lipton.

The Levers, however, have acquired control over another large manufacturer of soap and oil products, TOMCO. They have also taken over Kissan, a company known nation-wide for jams and squashes, and are reported to have acquired rice-milling facilities. Brooke Bond, an associate of the Levers, has entered marketing of *masalas*. Among the new entrants to the edible oil industry are ITC and Britannia. Parle, the market leader in the soft drinks segment, which had fought tooth and nail against the entry of Pepsi, was obliged to abandon its fight with TNCs and join hands with Coca-Cola. It appears that the withdrawal of restrictions on the use of foreign brand names has speeded up the process of domination of the Indian consumer goods market by transnational corporations. The entry of the U.S. based TNCs has coincided with the Indian policy to give high priority to private foreign direct investments and revision of the licensing policies to permit entry of large Indian companies and TNCs into the food processing industry.

The establishment of the new Ministry of Food Processing Industries (MFPI) at the Centre is an indication of the Government's thinking. The rationale given for the creation of the new Ministry is that the food industry has adequate market potential within India and a large scope for exports. Given the agenda set for the MFPI and the nature of enterprises, which are likely to be established under the new policies and the changed Industrial Policy environment, especially with regard to foreign private capital, the extent to which the sector;

- to successfully create a mode of operation and management in the food processing sector that would ensure increased incomes accruing directly to the producers, who are in the main concentrated in the rural areas;

- to create increased job opportunities in the rural areas with specific reference to women and unemployed youths by development of primary produce through a network of processing units in the various sectors;
- to bring the power of modern technology and marketing techniques in the aid of the farmers;
- to take the initiative in mobilizing cost effective technologies for storage, processing and marketing of agricultural produce;
- to think in terms of organizational restructuring of the domestic market so that overall demand is stimulated which, in turn, will lead to the growth of the food processing sector; and
- to ensure that adequate surpluses are created consistent with price and quality to further exports and earn valuable foreign exchange for the country by providing critical inputs to the industry to foster production for exports.

The general thrust of the attempts is to remove entry level restrictions. Major elements of the new policy provisions are:

- Industrial licensing requirement dispensed with except for 18 industries.
- Area reserved for the public sector curtailed so that it covers now only eight industries.
- Automatic approval for foreign direct investment up to 51 per cent of equity in certain high priority industries subject to balancing of dividend payments by export earnings over a period of time. This condition was later withdrawn except for certain consumer goods industries. The industries qualifying for automatic approval are: (a) all food processing industries other than milk food, malted foods and flour but excluding the items reserved for the small scale sector; (b) soya products, which include (i) soya texture proteins; (ii) soya protein isolates; (iii) soya protein concentrates; (iv) other specialized products of soyabean; and (v) winterized and deodourized refined soyabean oil; and (c) all items of packaging for food processing industry excluding the items reserved for the small scale sector.
- For foreign technology agreements, automatic approval is granted, provided the technology fee payments are under Rs 10 million and the royalty is within 5 per cent of domestic sales or 8 per cent of export sales, subject to a ceiling on total payments at 8 per cent of sales over a 10-year period.
- Use of foreign brand names allowed even for sales in the domestic market.
- Fast track approval, if the foreign equity covers the foreign exchange requirements for import of capital goods.

- Foreign majority trading houses allowed.
- Waiving of most of the conditions attached to letters of intent and industrial licences issued earlier.
- Easing of locational constraints.
- Abolition of phased manufacturing programme, which was intended to indigenize imported production process over a period of time.
- No specific permission needed for hiring foreign technicians.
- Foreign institutional investors allowed to invest in the Indian capital market.
- Mandatory convertibility clause associated with loans by financial institutions removed.
- Requirement for clearances for expansion and diversification under the monopoly legislation on the basis of size removed.
- Though reservation for the small scale sector continues, large houses and foreign companies are allowed to take up to 24 per cent of equity in companies owning small

Ministry's objectives would get pursued needs a closer scrutiny. The MFPI is obliged to create *increased job opportunities in the rural areas with specific reference to women, and unemployed youth by development of primary produce through a network of processing units.*

The MFPI is expected to promote modern technology and marketing techniques in aid of the farmers. The desirability of the modernization policy would depend on an evaluation whether the new technology would cause liquidation of the existing enterprises or it can be absorbed by smaller establishments to achieve higher productivity. Modern technology could help raise the average productivity in food processing, but to expect modern food processing industries to create a substantial rural job opportunities may not be realistic. If the intention is not to promote public sector or the cooperative sector, the alternative left would be to permit large industrial establishments, owned and controlled by Indian large business groups or TNCs, individually or in collaboration. Modern scale units.

- Except for distillation and brewing of alcoholic drinks, sugar and animal fats/oils and the areas reserved for small scale industries investments in the food processing sector are wholly exempt from any licensing obligation.
- Imports and exports freely allowed except for short negative lists.
- Procedure for capital goods imports simplified.
- De-canalization of a number of export and import items.
- Foreign exchange controls eased.

The observations of UNIDO are relevant in this context. In sum, "success" stories such as those of the poultry and cereal milling industries are, at the same time, failure stories. They are the symbols of the growing disarticulation of the agro-food systems in the developing countries, and of the growing dependence of these countries. Impressing growth rates in the two industries have led to accelerated marginalization of large sections of the rural population and a growing inability to rebuild an articulated agro-food system. This experience suggests that countries aspiring to create industries based on meat-poultry-cereal production should first consider developing their small-scale rural units, especially if they wish to build integrated national food systems that would make optimum use of local staples and generate employment and income in the rural areas.

In the context of the Industrial Policy followed particularly till the beginning of the 'eighties there appears to be a belief that growth of agro-processing industries was inhibited by the industrial licensing policy, in particular the reservation of certain products for production in the small scale sector. An examination of the licensing policy, however, would not warrant such a conclusion. If one goes by the reservation list one finds that very few items, most of which are in the nature of household preparations, fall under the reserved category. Even those initially reserved for the Small Scale Sector (SSS) have been over time de-reserved either partially or completely.

The slow growth of agro-processing, more specifically food processing industries, has been more due to lack of effective demand for the products processed and less for official restrictions. For instance, Cadbury (India) Ltd, an affiliate of Cadbury, U K, a British transnational corporation, established an apple juice concentrate plant in Jammu & Kashmir during the 'eighties. The venture did not meet with much success. Similarly, the capacity utilization at Mohan Meakins Ltd of juices and canned products was less than 40 per cent of the not very large automatic plants of large capacities with "not touched by hand" operations would add little to create new employment, on the one hand, and may, in practice, replace the traditional technologies and practices that were labour-intensive on the other. The employment potential of such processing units cannot, therefore, be large. Also, the employment may only be of seasonal nature. This being the scenario, it is but logical to fear that the objective of making a meaningful impact on the level of employment of women and rural youth is hard to achieve.

Food processing industry in the modern sector is likely to have its main characteristic in "brand" domination. The poor hygiene, in which the traditional food processing is generally undertaken, does not help sustain high confidence in the processed food produced and packaged by the cottage and small units. Bulk food processing is amenable to standardization and offers certain economies of scale. It is no surprise, if brand names get associated with quality and standardized products. The consumer acceptability of branded goods is high. It is because of such well known advantages that brand names carry substantial premium.

Emergence of near monopoly situation in food processing is not unknown. And once a brand name gets accepted widely, it opens up possibilities for launching new products with a massive thrust. Indian consumers are well aware how `Maggie' noodles were followed up by Maggie sauce and Maggie soups. It seems that, instead of offering more employment, the modern food processing plants can cause the closure of many cottage and small establishments. There is, however, a possibility that, because bulk food processing enterprises cannot provide large varieties or cater to regional and personal tastes, there would remain a demand for products of the cottage, local and small enterprises which provide specialized services with personal care. The overall impact of the entry of large and modern units to the food processing sector can, however, hardly be creation of promising employment opportunities.

Location of a large food processing unit would bring in a substantial increase in the demand for the required farm produce. This would also provide better market prices to the farmers. Given the fact that there would be thousands of small farmers who would be suppliers of farm produce to one buyer/processor, the market situation would invariably be of monopsony, the stronger one seeking to exploit the situation to its maximum advantage. The past experience in two important agro-based industries, viz sugar and tobacco, shows that the processing units tend to thrive at the cost of the farmers. The sugar industry of U P and Bihar is a case in point. On the other hand, whenever processing is undertaken in a cooperative framework, there is a faster growth both for the farmer and the industry.

The sugar cooperatives of Maharashtra, Gujarat and Karnataka have won a name for themselves. The same holds true of the edible oil (groundnut) cooperatives of Gujarat and the Amul experiment of dairy development on cooperative foundations. Thus, location of large private sector agro-processing units may be accompanied by perpetual conflicts, supply or processing uncertainties, labour problems, farmers' agitations and political interventions. On the other hand, if there were large farms specializing in a crop to meet requirements of the agro-processing unit, there could be a more organized and stable relationship between the farms and the factory. An important policy assumption is that agro-industries have a large potential for exports from India. One wonders, if at the present stage of India's level of food, vegetables and other farm production there is a scope for generating export surpluses. Can India be a net exporter of food in the foreseeable future? The per capita availability of cereals, pulses, vegetables and fruit in India is so low that export surpluses can only be obtained at the cost of their consumption. The poor would, of course, suffer the most. The early decades of this century witnessed foodgrain exports even when the average per capita availability of grains was extremely low and near famine conditions prevailed in many parts of India. The impact of exports of essential vegetables was well demonstrated by the abnormal price rise for onions during the mid-'seventies. In essence, the question is whether India should undertake food exports when millions remain half-fed. The argument here is not against strengthening the food processing industry; it is against the export argument for setting up the industry.

Like many policy assumptions, the export argument appears to be a misplaced one. For instance, a number of foreign collaborations — more so the financial ones — have been approved since 1986. With the beginning of the Seventh Five Year Plan, processed foods have been identified as one of the thrust sectors for exports. Yet in 1989-90, one finds very little of "processed" products in the export basket of agricultural and processed foods.

Very few items account for bulk of exports of agricultural and processed foods. Nearly one-third of the exports was accounted for by the fresh fruits and vegetables category, which includes onions. This category's share increased from 22.81 per cent in 1980-81 to 32.69 per cent in 1989-90. Indeed, onions had very high shares in both the years — 14.48 per cent and 17.77 per cent respectively. In this setting, one fears that the hopes associated with the establishment of the Ministry for Food Processing Industries both on employment as also on export front are likely to be belied.

In the foregoing, we have discussed how the Indian nationalist view on agro, rural, and cottage industries got crystallized during the pre-independence period and how the *Swadeshi* movement was directed to undermine the demand for cheap and imported machine made consumer goods, on the one hand, and to mobilize rural support for the struggle for India's political independence, on the other. We also reviewed, in brief, the role assigned to low-capital and labour-intensive technologies to maximize investment in basic and heavy industries under the Mahalanobis strategy.

The Third Plan onwards, however, the state has continued to provide support to traditional rural industries especially the *Khadi* and village industries. Agro-industries received impetus at two distinctively different phases. *One*, during the 'seventies when, as a consequence of the green revolution, State Agro-Industries Corporations were established to provide modern agricultural inputs: and, *two*, during the 'eighties, when the role of foreign direct investment and technology in the food processing sector was emphasised. It is yet not evident whether the latest approach would yield the desired results in the form of either increased employment opportunities, particularly for women and rural youth, or enhanced exports. On the other hand, there is a possibility that many small and local establishments may get adversely affected.

What is needed is a fresh and comprehensive approach, integrating the development of villages with agro-industries, with larger involvement of the farmers in processing their own produce. In a democratic set-up, one cannot ignore development of the majority of the people or keep them on subsidies. The fact also remains that Indian population is so distributed that migration from agriculture to industry or from rural to urban centres or from densely populated areas to scarcely populated ones is not an easy and sustainable alternative. Besides the physical dimensions involved, the very characteristics of the population are such that there are clear linguistic barriers, which limit large scale population migrations. Gainful employment to the rural people has to be provided in their own locale. Viewed in this perspective, agro-industries as a concept have to be dealt with very differently from the past approaches, policies and programmes or other industries.

In India, unfortunately, rural development, employment of the educated or the uneducated, equity, and phrases like social justice have not travelled beyond a pious demonstration of the good intentions of the state. Expressions such as these have found frequent entry into official policy announcements. The Indian Plan strategy was never operationalized for redistribution of wealth or for reduction of inter-personal or interregional disparities. Indian planning has, in the main, leaned more on investments than on the *policy* content.

Consequently, old colonial and feudal institutions, social value systems, educational structures and citizen-state relationships have never undergone any marked change. Rural programmes under the Five Year Plans, except land reforms during the 'fifties, though consolidated, were only extensions of the earlier individual rural development programmes and there was no specific programme designed for *area* development, where a region's resource endowment is sought to be optimally harnessed by relating Plan investment/ Plan programmes or pattern of allocation to the nature and the magnitude of local resources.

5

Agricultural Marketing in Worldwide

INTRODUCTION

This publication reviews the experiences of non-governmental and community-based organizations (NGOs and CBOs, respectively) in agricultural marketing initiatives. Many NGOs target the rural poor, whose livelihoods are generally focused on primary agriculture or trade, processing and services linked to the agricultural sector. The ability of those rural communities to access remunerative markets is a critical determinant of incomes and well-being. This publication examines the evidence on NGO and CBO agricultural marketing interventions in sub- Saharan Africa and, to a lesser extent, other developing regions, concentrating principally on access to domestic markets.

It highlights examples of best practice, explores the policy implications of those intervention strategies, and signals particular dilemmas or areas where further research is needed. This is one of a series of publications that seeks to amplify the relationship between poverty, rural livelihoods and key policy areas.

The publications are intended for a wide audience in developing country governments, donor agencies, research institutes and other organizations concerned with development or governance. They are intended to contribute to increased focus on poverty in development by informing and stimulating debate, policy and action amongst key players in the development process.

The publication is divided into four main sections:

1. This first section introduces the material to be covered and explains some of the issues and concepts that dominate the debate on agricultural marketing;
2. The second section explains why NGOs and CBOs are increasingly involved in this apparently commercial arena, and discusses some key differences in the orientation and role of different organizations;
3. The third section reviews different types of interventions, grouped in eight distinct categories;
4. The fourth and final section draws conclusions and points the way forward through policy and selective areas of further research.

MARKETING FUNCTIONS

In modern marketing the agricultural produce has to undergo a series of transfers or exchanges from one hand to another before it finally reaches the consumer.

This is achieved through three marketing functions:

1. Assembling,
2. Preparation for consumption
3. Distribution.

Concentration pertains to the operations concerned with the assembly and transport from the field to a common assembly field or market. The produce may be taken direct to the market or it may be stored on the farm or in the village for varying periods before its transport. It may be sold as obtained from the field or may be cleaned, graded, processed and packed either by the farmer or village merchant before it is taken to the market. Some of the processing is done not because consumers desire it, but because it is necessary for the conservation of quality.

At the market the produce may be sold by the farmer direct to the consumer or more usually through a commission agent or a broker. It may also be purchased by traders, wholesalers or retailers. The transactions may be carried out by direct negotiation or through middlemen, by barter or cash, by open or under cover auction, on the spot or in future markets. The transactions take place at one or more levels in the primary, secondary or terminal markets or all three. Distribution (dispersion) involves the operations of wholesaling and retailing at various points. By a series of indispensable adjustments and equalising functions, it is the task of the distribution system to match the available supplies with the existing demand.

MARKETING FUNCTIONARIES (AGENCIES)

The transfer of produce or goods takes place through a chain of middlemen or agencies. In the primary market the main functionaries are the producer, the village or itinerary merchant, pre-harvest contractors, commission agents, transport agents etc. In the secondary market the processing and manufacturing agents are the additional functionaries. Financing agents such as shroffs, banks and co-operatives may also take part. In the terminal or export market the commercial analyst and shipping agent also gets involved in the transfer of goods.

The functionaries have their own set up. They may be individuals, partners or co-operatives who may buy and sell on ready and future basis at a price determined by forces of supply and demand. Each functionary renders some service in the process of marketing and also earns a varying margin of profit for himself. This procedure makes marketing rather complicated and inflates the price of the produce. The nature of some of the agricultural products for example their bulk form and perishability and their seasonal availability further add to the complexity of agricultural marketing.

MARKETING IMPROVEMENTS

India being a primary producing country, agriculture plays a vital role, both as an essential infrastructure and a development component in generating and sustaining a higher national income. Out of a national income of about ₹38,921 crores in 1972-73 as much as ₹17,500 crores or about 44.9% is contributed by agriculture and allied sectors. It is estimated that about 50% of the agricultural produce is available as marketable surplus. The marketing system in India provides sustenance for about 3 million persons who are engaged in performing various marketing function. In the field of exports too, the agricultural sector accounts for about 50% of the total value.

The process involved in the disposal of such a substantial produce of great economic importance are significant not only for the farmer but also for the country as a whole. The unreasonably low return that the farmer gets for his produce and the excessive margin of profit retained by the intermediaries attracted the Government's attention and it was felt that the economic condition of the agriculturists could not be improved unless determined steps were taken to establish an orderly system of marketing in the country.

GOVERNMENT REGULATORY PROGRAMMES

With this object in view, a number of marketing surveys were conducted by the Directorate of Marketing and Inspection which revealed the shortcomings in the country's marketing system. A rectification of these deficiencies was sought to be achieved by rationalizing various activities and standardizing various practices in the markets through legislation or otherwise. The primary objective of improving the system of agricultural marketing was not only to remove the handicaps from which the producer-seller was suffering but also to increase his income by ensuring him a fair price.

REGULATION OF MARKET

Prevailing market practices and market charges made a deep cut in the share of the producer in the price paid by the consumer. Some of the market charges were authorized whereas others were more than what the service rendered warranted. It was felt that a remunerative price to the producer could only be ensured if the market practices and market charges were regulated and rationalised. And thus the regulation of the markets has been given a high priority in the various Five Year Plans. The markets are sought to be regulated through an Act of each legislature. The Act is generally known as the Agricultural Produce Markets Act and it is provided for the removal of various malpractices widely prevalent in the markets for the settlement of disputes between sellers and buyers and for the promoting of orderly marketing of farm produce in general. Various state Governments have made considerable progress in this field by bringing in the necessary legislation. The Acts enabling the respective states to regulate the markets generally provide for the notification of market areas and the commodities to be covered in the act in different areas. A 'Marketing Committee' consisting of the representatives of growers, traders, merchants, local bodies and Government nominees administers the working of each market. The functions of the market committee are to frame bye-laws, define local market prices, fix market prices payable to various functionaries, license the functionaries, settle disputes, supervise weightment and promote the development of orderly marketing in general. The committee is generally empowered to raise funds for its working by levying a small fee on the produce bought and sold in the market in addition to the license fees received from the functionaries.

Market Surveys

A survey conducted in 496 markets in 1961-62 has shown that after the regulation the market charges have been reduced by 48%. Another survey of selected commodities revealed that for some commodities market charges have been reduced by as much as 98%.

Although the first market to be regulated was Karanja in 1886 the regulation of markets did not make headway till the first Five Year plan. It got a fill up in the second and third five year plan.

The states and union territories which have regulated the markets are: Andhra Pradesh(335), Bihar(62), Chandigarh(1), Delhi(3), Gujarat(212), Mysore(102), Madhya Pradesh(233), Haryana(80), Kerala(6), Maharashtra(212), Orissa(34), Punjab(95), Rajasthan(90), Tamil Nadu(136), Tripura(1), Uttar Pradesh(264), Goa, Daman and Diu(1), West Bengal(15), and Himachal Pradesh(5). The states and union territories which are yet to regulate markets are Assam, Andaman and Nicobar islands, Arunachal Pradesh, Dadra and Nagar Haveli, Jammu and Kashmir, Laccadive and Minicoy islands, Meghalaya, Mizoram, Nagaland, and Pondicherry. Necessary measures to regulate markets in these states and union territories are at various stages of progress. It is expected that these states and union territories will regulate the markets by the end of the Fifth plan.

All the states where necessary legislation has since been passed, have formulated phased programmes for the regulation of markets. By the end of the fifth plan it is expected that all the wholesale assembling markets would be brought within the regulatory orbit.

As a result of this scheme, excessive commissions and other market charges have been substantially rationalized. Unauthorized and arbitrary deductions have been prohibited and malpractices stopped. The issue of sale slips by licensed commission agents to the sellers, indicating the details of sale proceeds, deductions effected etc. has been made obligatory. Weightment is also done by licensed weightmen of the market committee.

The dissemination of marketing information and news is one of the functions of the market committees. This is done through the displaying of the prices prevailing in the market and also in the neighbouring markets on the notice boards and announcements through loud speakers at regular intervals. This information is also supplied to the Central Government and the State Government and also to other market committees. Arrangements have been made for the maintenance of reliable statistics arrivals, sales, stocks, prices etc. which are maintained by the market committees.

It is also obligatory on the part of the market committee to provide the market yards with the necessary amenities. In some of the markets that have been regulated amenities like rest houses, cattle sheds and water troughs have been provided by the market committees for the convenience of the producer sellers. Facilities for grading before sale and storage have also been provided.

A survey of 500 regulated markets was undertaken by the Directorate of Marketing and Inspection in the Ministry of Agriculture in 1970-71 and 1971-72, with a view to assessing the adequacy and efficiency of the existing regulated markets and highlighting their drawbacks and deficiencies and suggesting measures to develop them. One of the most important drawbacks has been the inadequate financial resources of some of the market committees. During the fourth plan, a central sector scheme was drawn up by the Ministry of Agriculture to provide a grant at 20% of the cost of development of market, subject to a maximum of ₹ 2 lakhs. The balance will have to be provided by the commercial banks.

An important development in the field of regulated markets is the keen interest taken by the International Development Agency (IDA) in the development of the infrastructure in regulated markets. The IDA is financing the development of infrastructure in 50 markets of Bihar.

The World Bank has approved a loan assistance of 6.5 crores to Karnataka also for the development of markets.

CONTRACT TERMS

Under the existing trade practices, the sale of produce in a primary market takes place on the basis of the visual inspection of the goods, and in the secondary and terminal markets on the inspection of the samples. Thereafter the buyer and the seller decide upon the terms either orally or through written contracts. The contract terms specify the quality and quantity of the produce, the time and place of delivery, the price and terms of payment, handling and incidental charges, the procedure for settlement of disputes and penalties. The terms of contract were not standardized and thus varied for every individual transaction, and were more favourable to the buyer. With a view to improving trade practices, All-India standard contract terms have been drawn up for a number of commodities. In standard contract terms the definition of quality and allowances in respect of refraction, damaged goods have been specifically standardized though the adoption of these standard contract terms by traders is voluntary they have to a large extent strengthened the position of the producer-seller and have improved the quality of the product marketed.

STANDARDIZATION AND GRADING

In order to gain the confidence and establish a rational relationship between the quality of a produce and its price, it is necessary to devote some attention to the proper preparation sifting and sorting of a material according to certain attributes before it is taken to the market. This is sought to be achieved by grading the produce in conformity with certain accepted quality standards *viz.* shape, size, form, weight, and other physical and technical characteristics. The produce brought to the market is very often contaminated with dust, stones and other foreign matter added either deliberately or by accident. Sometimes the produce is immature or not properly dried or contains shrivelled grains or damaged and rotten material. Such a produce brings a lower price to the farmers. Care should be exercised while assembling the produce of different farmers so that the good material is not mixed with the inferior material brought in by some farmers.

The Government of India had recognized the need to introduce the standardization of agricultural produce which would enable the farmers to derive the benefits of grading in terms of fair practices according to the prescribed standards and enacted the Agricultural Produce Grading and Marking Act in 1937. The Act empowers the central Government to prescribe grade standards indicating the quality of articles included in the schedule and specify grade designation marks to represent particular grades or qualities. The Act provides for the grading and marketing of agricultural produce. The grade standards prescribed under this act are based on both physical and chemical characteristics and are formulated after analysing representative samples of each commodity collected from different regions and different seasons. Besides the international standards and special requirements of overseas consumers are also taken into account while formulating these standards for the commodities which are exported. The grade standards are reviewed and amended from time to time in the light of the shift of the pattern of production and trade and changes in the consumer's preferences. The grades are designated as the 'Agmark' grades.

A central Agmark Laboratory at Nagpur with sixteen regional laboratories at Guntur, Madras, Bombay, Kanpur, Cochin, Rajkot, Calcutta, Sahidabad, Jamnagar, Bangalore, Patna, Tuticorin, Virudhunagar, Mangalore, Alleppey and Kozhikode are assisting to provide adequate laboratory facilities for fixing grade standards for new commodities, for revising old grade standards and for routine quality control work.

Grade for Export

Grading of agricultural produce under the A.P. (G and M) Act is voluntary. Exports of certain agricultural commodities have however been prohibited unless duly graded and marked in accordance with the grade standards laid down under the A.P. Act 1937. The power to so prohibit exports was derived under the provisions of the sea-customs act.

The Directorate of Marketing and Inspection under the Ministry of Agriculture exercises a three tier control on the quality of agricultural commodities that are graded under 'Agmark' before they are exported. This is done through inspection.

Grading for Internal Trade

Commodities such as cotton, ghee, butter, rice, wheat, atta, gur, eggs, arecanut, potatoes, fruits, bura, pulses, vegetable oils and ground spices are being presently graded under 'Agmark' on voluntary basis.

Grading at Farmers Level

The grading of agricultural commodities under 'Agmark' has been consumer oriented. Generally the grading was done at the level of the traders. At this stage the producer was not a direct beneficiary of the grading scheme. It was felt the need to introduce grading at the producers' level. Thus the Directorate of marketing and inspection introduced a scheme for setting up commercial grading units.

Grading of Fruits and Vegetable Products

With a view to exercising quality control over fruits and vegetables the Government promulgated the Fruits Product Order under the essential Commodities Act. The preservatives and colors to be used are also clearly laid down. The order also stipulates the hygienic and sanitary methods which must be adopted by the manufacturers. The license for all this is issued by the executive Director, food and Nutrition board.

The total number of licensed factories was 1194. The total average production of fruits and vegetable products in the country during 1970 has been estimated as 156 thousand tonnes valued at ₹32.66 crores. Some of the products are very popular in foreign markets and are a good source of foreign exchange. In 1970 alone fruit products worth ₹2.92 crores were exported from India.

Regulation of Cold Stores

Most of the problems relating to the marketing of fruits and vegetables can be traced to their perishability. Perishability is responsible for high marketing costs, market gluts, price fluctuations and other similar problems. At low temperature, perishability is considerably reduced and the shelf life is increased and thus the importance of cold storage or refrigeration.

The first cold store in India was reported to have been established in Calcutta in 1892. However significant progress in the expansion of the cold storage industry in the country has been made only after independence.

An ad-hoc survey of the cold stores carried out by the directorate of marketing and inspection in 1955 showed that the total available cold storage in the country was only 77 thousand tonnes. The survey also highlighted the need to regulate the cold storage industry in a planned manner.

With a view to ensuring the observance of proper conditions in the cold stores and to providing for development of the industry in a scientific manner, the Government of India and the ministry of agriculture promulgated an order known as "cold storage order, 1964" under Section 3 of the Essential Commodities Act, 1955. The order is applicable in respect of every cold store with a capacity more than 8.4 cu m. The jurisdiction of this Order extends to the whole of India except West Bengal. Under this order, it is obligatory on every operator of the cold store to obtain a license from the Licensing Officer before using the installation for storing any food stuffs *e.g.* fruit, vegetables, meat, fish, dairy products. The Agricultural Marketing Advisor to the Government of India is the Licensing Officer.

The directorate of marketing and inspection is enforcing the cold storage order. The field staff posted in the regional and sub offices located in different states regularly inspects the cold stores and offers necessary guidance for better and scientific preservation of foodstuffs. Besides the directorate of marketing and inspection gives general guidance on all technical matters concerning the setting up of cold stores to the intending entrepreneurs.

The Government of India constituted a Central Cold Storage Advisory Committee consisting of official and non official members, representing the growers, owners, machinery manufacturers, research organizations etc. The Committee advises the Government on all matters pertaining to the enforcement of Cold Storage Order and the future development of the industry. At the end of 1973-74 there were 1503 cold stores in the country with a capacity of 18.70 lakh tonnes. Uttar Pradesh had the maximum number of cold stores of 9.37 lakh tonnes followed by Bihar with 169 cold stores with 2.10 lakh tonnes capacity and West Bengal with 133 cold stores with 3.15 lakh tonnes capacity. Out of 1217 cold stores licensed during 1970, 1021 representing 84% of the total were owned by the private sector, whereas 121 and 75 were owned by the public and co-operative sector respectively. The respective capacity is 13.88 lakh tonnes, 0.25 lakh tonnes and 0.73 lakh tonnes. Though numerically the cold stores in the co-operative sector were less than those in the public sector the capacity was more. The National Co-operative Development Corporation

(NCDC) has formulated a scheme for financial assistance for setting up new cold stores in the co-operative sector.

Potato is the most important commodity which is presently placed in the cold stores accounting for as much as 92% of the total capacity in the country. The remaining 8% is being utilized for other perishables *viz*. fruits, vegetables, meat, sea-foods, dairy and poultry products.

With the attainment of self-sufficiency in the production of food grains, greater attention is being paid to the increased production of protective foods such as fruits, vegetables, fish, poultry and dairy products. These products being highly perishable are required to be kept in cold stores. Owing to the inadequate cold storage facilities at present considerable losses occur in the case of these commodities. There is enough scope for expanding the cold-storage industry with a view to providing facilities for preserving and prolonging the shelf life of these protective foods which are essential for human health.

Consumer Protection

In the marketing process the producers and consumers are the two weak ends of the chain. It is incumbent on the part of the Government to protect the interests of both of them. Producers are sought to be protected through the regulation of the markets, grading at the producers level and other similar measures. Consumer's interests on the other hand are safeguarded by grading under 'Agmark' at the level of traders. The progress towards making the consumer quality conscious is slow. With the growing popularity of semi-processed foods the danger of sub-standard food articles being marketed has become manifold. With this in view steps have been taken in many directions, *e.g.* Acts relating to grading and standardization of agricultural commodities, certification marks in respect of manufactured goods, pure food laws, laws relating to weights and measures, and laws relating to the manufacture of fruits and vegetable products have been passed and enforced. Commodities such as ghee, vegetable oils, butter, honey and powdered spices are being graded and marked under 'Agmark' under the provisions of the Agricultural Produce (Grading and Marking) Act 1937. The Agmark attempts to provide a third party guarantee for the consumer. It not only certifies the purity of the product but also gives an indication of their quality by the grade-designating mark. The economic incentive to the producer-manufacturer is reflected in the premium the Agmarked product fetches in the market over the ungraded products.

Running parallel to the enforcement of laws, certain measures to complement the effort of achieving the overall objective of consumer protection are also being adopted. Monopoly procurement of food grains being operated by the Food corporation of India and other state departments achieves the two objectives of ensuring a remunerative price to the farmers and a uniform and reasonable price commensurate with the quality of the produce to the consumers. The task of consumer education and protection is formidable and the Government machinery alone will be inadequate to accomplish this task. There is a need of active consumers' movement as in other countries. The Consumers' Guidance Society has been established with the objective of educating the consumers about their rights and responsibilities and advising them on the quality of products available in the market. Similarly it has advised the producers and manufacturers to abide by such standards as are necessary for the health and safety of the users.

Co-operative Marketing

The existing institutional structure of co-operative marketing is such that the co-operatives are functioning at the primary level, at the secondary level (taluka or district) and at the state level. In pursuance of a recommendation of the Dantwala Committee, efforts have been made to persuade the state Governments to divert the middle tier *viz.* the district marketing federations, of the functions legitimately falling within the purview of the state or primary marketing societies, so that a two-tier system can be brought into operation. The co-operatives in different states have been federated into a central level federation *viz.* the National Agricultural Co-operative Marketing Federation (NAFED). At the end of 1960-70 as many as 3335 primary cooperative marketing societies were operating. Of these more than 500 were specialized commodity marketing societies for special crops, cotton, fruits and vegetables. In certain states intermediate organizations at district and state levels have also been established. At the end of 1960-70, 232 such societies were functioning. They included some commodity federations also. At the state level 28 apex co-operative marketing federations are functioning. They are normally handling all the commodities. There are a few apex cooperative societies which are handling exclusively a particular commodity *e.g.* two apex societies in Gujarat are handling cotton, one is handling fruits and vegetables and one in Uttar Pradesh is handling sugarcane only. In order to strengthen and develop co-operative marketing to an extent where it may have an impact on the marketing of agricultural produce, several measures have been initiated. Steps are also being taken to see that there is an effective co-ordination between the state cooperative departments and their counterparts dealing with agricultural marketing. Consequently there has been a significant expansion in the operations of marketing co-operatives. This is exclusive of the value of agricultural requisites and consumer articles handled by the marketing cooperatives. The main commodities marketed by the cooperatives were sugarcane, cotton, oilseeds, fruits, vegetables and plantation crops.

In pursuance of the decision of the Government, a scheme of outright purchases of agricultural produce by the co-operative marketing societies was launched in 1964-65 in 200 selected marketing societies. The basic theme of this scheme has been to bring the small producers within the fold of co-operative marketing together for such farmers. To provide the necessary financial backing the scheme envisaged the creation of a price-fluctuation fund. The fund envisages meeting losses if suffered by the co-operative marketing societies at different levels as a result of outright purchases of agricultural produce.

The scheme has been operative in several states. During 1969-70 the value of the agricultural produce purchased under the outright purchase scheme was about 34 crores. As a result of this scheme there has been a greater involvement of cooperative societies in the marketing of agricultural produce.

Besides this scheme has given impetus to inter-state trade by the cooperatives. They transacted about ₹ 66.55 crores worth of agricultural produce during 1969-70.

Interstate Trade

The co-operative marketing societies are devoting increasing attention to interstate trade in agricultural produce. The main commodities are wheat, pulses, plantation crops, copra, spices, fruits and vegetables. The bulk of the transactions were made by the Punjab Apex Federation. During 1970-71 the NAFED acted as the agency of the co-operatives of Jammu and Kashmir for marketing their apples outside the state.

Co-operative Export of Agricultural Produce

The export of agricultural produce by the co-operative sector continues to be a growing activity. The bulk of the exports are made by National Agricultural Cooperative Marketing Federation Ltd., which accounted for exports worth ₹5.64 crores. Besides, the Gujarat State Cooperative Marketing Federation Ltd., the Khanna Cooperative Marketing Federation Ltd., the Kerala State Cooperative Marketing Federation Ltd., the Jalgaon district fruit sales societies and the coconut oil millers society also cooperated to exports to countries like Kuwait, Malaysia, Ceylon, Singapore, Bahrain, Doha, Dubai, Iran, Muscat. The Khanna Solvent Extraction Plant in the Punjab State exported de-oiled cake worth Rs. 35 lakhs during 1969-70. The main commodities exported by the cooperatives were pulses, chillies, onions, pepper, de-oiled cake, potatoes and kardi extraction meal.

The exports were mainly made to Ceylon and other important markets were Mauritius, Kuwait, Doha, Bahrain, Hong Kong, the USSR, the UK, Iran, Czechoslovakia and France. Some of the traditional items of export have been marketed in non traditional areas. Pulses were exported to Cuba and onions to Malaysia and Singapore. The cooperatives also assisted various agencies to export agricultural commodities. In this connection exports of coffee and raw sugar were made. During 1970-71, NAFED exported agricultural produce worth ₹5.26 crores. The Jalgaon District Fruit Sale Societies Cooperative Marketing

Federation Ltd., directly exported bananas worth ₹34 lakhs to Kuwait and Bahrain Islands.

Co-operative Cold Stores

Co-operatives are also to facilitate the storage and marketing of perishable commodities especially seed potatoes. By the end of the Third plan the cooperatives had established 87 cold stores and the target for the fourth plan was set at 45 more. At the end of December 1971 there were as many as 96 cooperative cold stores with a capacity of 1.42 lakhs tonnes.

TRAINING OF MARKETING PERSONNEL

Several training courses, viz.:

- 11 months Diploma Course in Agricultural Marketing in Nagpur,
- 4 months Market Secretaries Course at Chandigarh, Lucknow and Hyderabad,
- 6 months Diploma in Livestock Marketing at Nagpur,
- 3 months training for grading supervisors at Nagpur
- 3 months Graders Course at Hubli, Lucknow, and Chandigarh,
- 4 months Training Course in Cotton Classing Centre at Surat,
- 6 months Training Course in Tobacco Grading at Guntur,
- 3 weeks course at Demonstration cum training centre in animal casing at Bombay and New Delhi are being run by the directorate of marketing and inspection.

In addition short term condensed training courses of one week duration organized for the state Government personnel in Kapas grading. As many as 407 candidates in the 11 months Diploma course, 1596 in 4 mths course; 212 in 3 mths course; and 76 in 3 mths grading course were trained by the end of 1972-73.

Market Extension

The directorate of marketing and inspection has set up a separate extension wing for the dissemination of information valuable to producers as well as to consumers. The primary aim of the scheme is to enlighten the producer seller on consumer preference and to advise him on the proper methods of preparation for marketing, grading, storing, packaging, handling and transporting and to improve the quality of the produce and to secure a better return to the growers.

Since 1969, the Directorate of marketing and inspection has in collaboration with the directorate of extension launched a series of 'Agmark' exhibitions known as 'vital link between farm and home'. These have been held in New Delhi, Lucknow, Chandigarh, Trivandrum, Bombay, Calcutta, Bangalore, Nagpur, Ahmedabad, Madras and Hyderabad. These have designed to educate the general public in the field of agricultural marketing to promote quality consciousness among producers.

Marketing Intelligence

With a view to disseminating the marketing intelligence to the interested parties the directorate of marketing and inspection publishes the following journals:

1. Agricultural Marketing—a quarterly journal
2. Marketing Newsletter—a monthly letter
3. Agmark Statistics—yearly
4. Commodity Intelligence Bulletins for tobacco, wool, bristles, potatoes etc.

MARKETING RESEARCH AND INVESTIGATION

The role of market research in the establishment of an efficient system of marketing cannot be overemphasized. In order to be able to introduce reforms one should know the defects and shortcomings of the prevailing system. This calls for and justifies the necessity of intensive research and investigation. Under the third five-year plan an elaborate scheme for conducting survey, and research in the various facets of agricultural marketing has been undertaken by the Directorate of marketing and inspection. Particular attention is being paid to the collection of market information, with regard to price spreads, shifts in marketing practices, consumption pattern, consumer preferences, directional movements, packaging, assembling, transportation, distribution etc. Authentic statistics and data are being collected so that up-to date information may be maintained in respect of all important agricultural commodities. Besides the market surveys cover studies on the organizational aspect of the marketing system, problem oriented studies etc. Under the fourth five year plan a specialized study to estimate the post harvest losses and marketable surpluses of agricultural produce have been taken up on an all-India level.

The Research Wing of the Directorate of marketing and inspection has been further strengthened by creating the Market Research and Planning Cell. The cell has a big component of exports in the field of agricultural marketing research and is expected to increase the tempo of development in this field.

Marketing is a multistage process. For the improvement and development of marketing structure, a co-ordinate approach aiming at removing all the weak links of the marketing chain is essential. A package of improved marketing services in the form of regulated markets, grading, weighing, storing, transporting, handling services and marketing finance need to be made available to ensure the producer a fair return from his production efforts and a better share in the price paid by the consumer. At the same time, market research programmes should be oriented to the developing of an orderly and efficient marketing system. This is a crucial time in the development of agricultural marketing when the country is poised to enter an era of production surpluses. A piecemeal approach at this stage can be disastrous and can nullify the advantages gained by the farmer on the production front.

AGRICULTURAL MARKETING IN DEVELOPING COUNTRIES

Economic reforms have had sweeping impacts on agricultural markets in developing countries.

In general, state intervention has been reduced, notably with respect to:

- The abolition or sharp curtailing of parasitical marketing boards;
- Depreciation of formerly over-valued currencies rendering developing country exports more competitive and imports more expensive;
- A reduced public role in agricultural services, especially in subsidized credit, input and extension networks;
- A shift away from pan-territorial and pan-seasonal crop pricing strategies and pre-announced prices.

Reviewing agricultural markets research in sub-Saharan Africa and Asia, Jones concludes:

- "In newly liberalized [food] markets in eastern and southern Africa... barriers of entry to trade are low, but the marketing system has little capacity to channel credit or spread risk. There are strong theoretical reasons for expecting the impact of and response to reforms to vary between different classes of producers. The absence of key markets, risk aversion, high transaction costs and the dual role of agricultural households as producers and consumers are critical features. The marketing system depends on both physical and institutional infrastructure... Collective action by market participants may address this but it may also lead to collusion over prices. Evidence from South Asia shows that food markets exhibit social barriers to entry, massive asset polarization, debt relationships between large and small traders and traders and farmers, diverse institutional and contractual arrangements, and collusive behaviour, enforced in part by manipulation of the state regulatory system."

This conclusion gives some clue to the reasons why NGOs and CBOs intervene in agricultural markets. When extension agents, researchers and development organizations working in rural areas ask farmers to prioritize their problems, agricultural marketing is repeatedly raised as one of the most important problems faced. It may arise in the context of the promotion of new crops or productivity-enhancing technology, or it may be felt particularly acutely in remoter areas poorly served by commercial traders, where parastatals no longer operate. NGO marketing interventions typically aim to fill critical gaps in the marketing system or address the power imbalances to which Jones refers. Nowhere are those marketing problems felt more acutely than in the areas for which it is most difficult to identify sustainable strategies to improve market access. Farmers in remote areas (either remote because of physical distance from markets or because of poor roads) are almost always poorly served by agricultural traders and are often obliged to accept seemingly unattractive prices for their produce. Distance from markets rules out the production of higher value more perishable crops, and reduces the linkages between these producers and other more specialized markets.

By the same token, CBOs and NGOs seeking to promote alternative strategies for these disadvantaged communities face high costs and tangible obstacles that make their task particularly difficult. Poor access to markets is mirrored by poor access to all kinds of rural services. The poverty that results makes such communities particularly risk-averse. Where rainfall is uncertain, the situation is even worse, whilst the relative absence of trade does nothing to relieve the covariance in production. These are the challenging circumstances that make an examination of marketing interventions worthwhile. There is wide-ranging experience amongst the development NGO community. Some of these initiatives have taken-off and developed into self-sustaining activities, whilst others, although not conceived as such, have effectively become subsidy dependent welfare programmes. This review identifies best practice and the conditions required for such programmes to work.

NGOS AND CBOS - SOME DEFINITIONS

NGOs are part of the development landscape. Increasing amounts of development aid are channelled through NGOs.

The term gives little clue as to their real characteristics but most people associate NGOs with the following:

- A formal and officially recognized organization that is not linked to government;
- Having a purpose that is altruistic rather than commercial;
- Attracting staff who are value-driven rather than financially motivated.

Fowler highlights some features of the voluntary sector by making comparisons with government and business- organizations.

Table. Comparison of Organisations in Different Sectors

	Sector		
Characteristics	**Government**	**Business**	**Voluntary**
Relationship to those served based on:	Mutual obligation	Financial transaction	Personal commitment
Duration of relationship to those served:	Permanent	Momentary	Temporary

Approach to external environment:	Control and authority	Conditioning and isolation	Negotiation and integration
Resources from:	Citizens	Customers	Donors
Feedback on performance	(in) direct politics	Direct from market indicators	"constructed" from multiple users

The term 'community-based organization' or CBO may be associated with similar values but is generally more focused on issues particular relevant to the community from which its membership is drawn. CBOs may be quite formally structured, but can equally be loosely structured, informal organizations; farmers' associations are an example of a CBO.

Stocker and Barbor-Might discuss CBOs in relation to civil society organizations (CSOs): "CSOs are, simply, organizations operating in civil society somewhere between the informal associational world of family, kin, neighbours, friends and the more formal or market-oriented world of business organizations, state – and NGOs. Although in practice there is diversity and complexity, organizationally the idea is a simple one; NGOs act as intermediaries for CSOs...

CSOs sometimes evolve into NGOs. Many of them have nothing to do with development or are not poorer people's organizations. With this in mind, many authors and authorities prefer to use alternative terms, for example, 'GRO' (grass roots organization) or 'CBO' (community-based organizations) to designate CSOs that exist to serve their members, these members being poorer people. CBO is the usage followed by the World Bank."

DIFFERENT TYPES OF ORGANIZATION

Although some definitions were provided in the previous section, these were not especially helpful in distinguishing between the plethora of organizations present in many developing countries.

A four-way categorization is proposed here, based largely on origins and capacity:

1. Northern NGOs with offices in developing countries, usually obtaining funds from donors (including private individuals); this group is quite broad since it encompasses large NGOs such as Oxfam or CARE, with activities in many countries, as well as small NGOs whose activities may be quite focused on a few countries and issues.
2. Indigenous NGOs who have become relatively large, well-organized and able to attract significant funds from international donors and Northern NGOs; sometimes these NGOs were originally created or strengthened by Northern NGOs.
3. Indigenous NGOs that are small, usually focused on a particular geographical area or issue, that obtain small amounts of funding from donors or government, but who struggle to grow or stay afloat.

4. CBOs, membership organizations serving particular interest groups usually in rural communities, whose focus may be broad or quite narrow; these organizations may be formally structured or quite informal; farmers associations, credit groups, and joint marketing societies could all be considered CBOs in the context of this review.

Many countries have laws governing the registration of different types of organizations that may confer a certain tax status or legal standing. Some developing countries have umbrella associations for NGOs. In any particular country it is useful to find out whether an umbrella organization exists, and if so, the types of CBOs and NGOs that tend to be officially registered – recognizing the potential divergence between official requirements and practice.

THE EVOLVING ROLE OF NGOS AND CBOS IN DEVELOPMENT ASSISTANCE

In the last 20 years, NGOs have become progressively more involved in development assistance, at every level. The shift from a relief and welfare focus has come about partly in an attempt to address the underlying causes of some of those man-made disasters or to limit the negative consequences of the natural disasters at which they assisted. It has been helped by the increased funding they found they were able to attract. Many Northern NGOs now have policy and research departments, and are a legitimate channel for large amounts of donor funding. At the same time, the role of the state has been redrawn, and in developed and developing countries, there is now a much greater focus on civil society as a way to improve democratic processes and bring about greater accountability in government. Governments are also seeking ways to be smaller and to sub-contract functions where feasible. Furthermore, funding developing country organizations to carry out development work is considered a way to build indigenous capacity. NGOs working in developing countries have benefited from this trend – either because they are considered part of civil society or because they work closely with many civil society organizations, including CBOs. As Stocker and Barbor-Might state:

- "From the point of view of the donors, civil society was the 'place' where something could be done and, often enough, NGOs were the intermediary institutions or midwives of such remedial programmes [relating to structural adjustment], spanning the gap between donors and CSOs. The funding channels varied, sometimes being directed through Northern NGOs (which might provide 'aid' directly or channel it to one or more partner Southern NGOs or CSOs) and sometimes going as direct funding to Southern NGOs and in some instances even to CSOs. When governments were irredeemably corrupt or oppressive (as, for instance, in Haiti during the Duvalier regime), these programmes seemed to offer virtually the only hope of channelling assistance to the people who most needed it."

This growth in the funding, remit, competencies and responsibilities of NGOs has meant that they have been closely involved in (if not the instigators of) much of the experimentation with practical solutions to pressing problems in rural areas. This is the context in which NGO experiences with agricultural marketing interventions provide a valid and rich focus for this review.

NGOS AND CBOS

Although many NGOs share similar altruistic goals, their approaches vary enormously. This is particularly evident in the extent to which they embrace and harness commercial activities to promote broader objectives, or reject this as a legitimate means by which to achieve social objectives. Moreover, amongst those NGOs prepared to use commercial activities as a means to an end, there can be considerable variability in the role these activities are accorded within thc development strategy and the competence with which they are planned and undertaken.

Organizations that are Primarily Welfare-oriented

A large number of NGOs and CBOs become involved in agricultural marketing activities, but this is rarely their core business. (There are some notable exceptions amongst some of the international NGOs who have become very experienced in agricultural enterprise, agro-processing and marketing. These include, for instance, TechnoServe, the Intermediate Technology Development Group, Enterprise Works Worldwide and the Cooperative League of the USA.) Many NGOs start with welfare (or social or altruistic) objectives, in areas such as education, health, water, infrastructure and agriculture and gradually shift towards a longer-term development focus. With this shift, small business and income-generation activities take on a greater role. Gibson describes this gradual transformation in terms of a continuum of different actions and attitudes.

Table. NGOs' Evolutionary Path in the Development of small Businesses and Income Generation

From ←	→ To
Relief and welfare	Development
Short-term	Long-term
Ideological	Pragmatic
Community-focused	Individual-focused
Targeted	Self-selecting
Grants	Market interest rates
Amateurish	Professional
Income generation	Small business
Social/ Technical	Economic/ Business
Instinctive	Strategic
Beneficiaries	Clients

Often NGOs and CBOs deliberately work in remote and disadvantaged communities and target the poorest households or individuals. These conditions, in combination with a general relief and welfare orientation, influence the strategies they adopt to achieve their objectives. For instance, direct or indirect subsidies may be used to improve access to markets (*e.g.* through the provision of transport, credit or inputs). An example of a direct subsidy is free or below cost use of transport (calculated on the basis of costs of fuel and driver and perhaps some portion of the vehicle costs). An indirect subsidy might involve charging a commercial (or break-even) rate on the vehicle hire but taking no account of the staff costs of implementing and managing the scheme. Whilst few people would suggest that the subsidy could continue indefinitely, there may be little consideration of how these activities can eventually be shifted to a more sustainable basis. The result is often that the programme attracts participation because of the subsidies, and once it ends there is little enduring impact.

Yet in the short-run these types of activities are attractive to NGOs because they have fairly immediate and visible (if not enduring) impacts and can (with varying degrees of success) be targeted to particularly disadvantaged groups (such as the poorest households, women, refugees, the handicapped or other marginalized social groups). An approach that seeks to use commercial channels may take much longer to develop and may place the intended target group at a disadvantage relative to other members of the community. Even when NGOs and CBOs do not intentionally adopt a non-commercial approach, market-oriented interventions are often subordinate to their core business.

This affects the way they are developed and managed, as well as the way they are perceived within and outside the organization.

Stanton gives four reasons for this lower status:

- Ideological – "...the core work assists those who most need it, income-generating work assists those who will exploit it to the greatest economic advantage";
- Resource allocation – reflecting the ideological perspective;
- Management – "the most willing volunteer for the job rather than external recruitment of people qualified and experienced in business management";
- Finance – these activities may be quite costly to implement and effectively monitor.

Furthermore, marketing activities are often managed and evaluated in the same way as other development activities, with insufficient attention to budgeting and profitability. A survey of income-generating programmes carried out by a range of indigenous and international NGOs in Welaita, south-west Ethiopia, found that little attention was given to the income generated and the profitability of different schemes. However, NGOs using subsidies to target disadvantaged groups would argue that this is a legitimate way to improve the livelihoods of poor individuals, households and communities, particularly in remote areas. Yet even with these subsidies, it may be difficult to have much impact on livelihoods in the most geographically and socio-economically disadvantaged communities. Stanton points out the disadvantages of this type of approach: the frequent failure to make a significant profit, high costs that prevent the NGOs/CBOs from reaching a wide audience, and problems concerning long-term sustainability.

Business-like NGOs and CBOs

- "Organizations which themselves resemble small businesses – in terms of their people, culture, systems, structure and behaviour – are most likely to be successful in encouraging the growth of small businesses."

In recent years, private sector development has increasingly been seen as a viablc and important approach to sustainable development. Thus many governments, NGOs and CBOs have focused on the promotion of marketing and small-scale enterprise to encourage greater participation in the commercial sector, as a route to higher incomes, employment generation and growth.

Small enterprise development work, which grew considerably in the 1980s, has contributed to a realization that it is possible to make much greater use of market mechanisms in pursuit of development objectives. This has been shown particularly through the success of micro-credit initiatives, where even very poor individuals are able to repay not only loans but also interest which sometimes covers the costs of providing credit. Also, social objectives and commercial objectives are not mutually exclusive and many NGOs and CBOs pursue both.

The fair-trade movement is a good example of this. Fair-trade organizations use commercial methods to generate social development benefits through improved terms of trade. The important thing is to balance potential marketing success with the social benefit needs of the beneficiaries. Furthermore, selective use of subsidies can still lead to sustainable and successful marketing initiatives, depending on the circumstances, as demonstrated by the CARE Egypt Agricultural Reform Programme. The programme provides information services to smallholder farmers and facilitates linkages to help increase farmer income.

The service is highly subsidized but has proven successful for a number of reasons:

- It helps to link farmers to sources of information outside the programme, thereby fostering the long-term sustainability of relationships and networks;
- Farmers contribute financially, *i.e.* they pay fees for the services;
- The demand for services is farmer-driven and project staff work with farmers to identify production and marketing opportunities.

A similar approach is used by Intermediate Technology in Zimbabwe. Assistance in product development is offered but the initiative must come from an existing business, which must be willing to contribute to the costs of product development (*e.g.* through materials, labour or workshop facilities). The two key advantages of greater commercial orientation and awareness are that cost-recovery enables more people to be reached by such programmes, and sustainability becomes a realistic goal. Gibson argues that NGOs appear to have distinct advantages in pursuing income-generation programmes ("smaller, more flexible, innovative organizations"). His conclusions are firmly rooted in the belief that commercial strategies can serve development objectives:

- "The continuing challenge is to progress from this base so that the economic growth of other developing countries is enhanced, is driven by indigenously owned and indigenously managed enterprises, and reaches the poor and disadvantaged sections of the population."

DIRECT INTERVENTION OR FACILITATION

The marketing role that NGOs and CBOs take on lies somewhere along a continuum between being directly responsible for marketing activities to facilitating beneficiaries/clients to market for themselves.

Direct Marketing Role

The term 'income-generating programme' (IGP) is used to describe a variety of programmes. These range from enterprises owned and managed by the beneficiaries to enterprises owned and managed by the organization which employs the beneficiaries. A number of NGOs/CBOs have established this latter type of small business to generate income to finance their other programmes and reduce donor dependence. CBOs and NGOs can also be more directly responsible for marketing activities. One way of doing this is through out-grower schemes (sometimes referred to as contract farming or satellite production). Such schemes involve smallholder producers providing agricultural raw materials to trading or processing businesses. Often growers work as a group, characterized by String fellow *et al.* as linkage-dependent groups. Generally the marketing arrangements are predetermined: prices or a pricing formula are agreed.

They help markets function to the benefit of both producers and the companies or organizations involved. This type of relationship is beneficial for farmers because they have a secure market for their produce at a predetermined price and the buyer benefits from having a guaranteed source of raw materials and lower transaction costs, which reduces his/her risk and costs. Within the fair-trade arena, the role of, and the marketing channels used by NGOs and CBOs (or alternative trading organizations – ATOs), also varies. Some organizations (such as Oxfam Trading and Traidcraft) take a direct marketing role by acting as wholesalers, with the producers acting as subcontractors producing to order. An advantage of this type of arrangement for producers is that they are guaranteed a volume of sales, thereby minimizing their risk. A disadvantage of this, and of out-grower schemes, is that the producers can be dependent on the trader, and may not have access to alternative buyers or markets if for any reason the trader is no longer able to market their produce.

ADC and Bean Outgrower Schemes in Uganda

The Agri-business Development Centre (ADC) in Uganda helped to establish bean outgrower schemes in Kasese and Kibaale districts together with the Uganda National Farmers Association (UNFA) and Bugangaizi Export Commodities Limited (BEC). The purpose of the schemes was to integrate poor rural farmers into the bean market, through marketing agencies, to provide them with an additional income source. ADC's role was facilitating the supply of bean seed and training some of the producers as 'farmer-extensionists' to provide follow-up extension advice to other producers. UNFA and BEC promoted and implemented the programmes. They were responsible for managing seed supply, training and extension and organized marketing (procurement and collection) of the crop.

The outgrower schemes have been successful from the producers' perspective in several ways:

- The numbers of farmers reached has steadily increased;
- Output levels have increased (participation, acreage and yields have all increased);
- An effective farmer-extension system has been created, and the adoption of new varieties and improved production methods has been good;
- Planting materials have been maintained and expanded through seed multiplication;
- Household income from bean sales has increased.

From the trader perspective, a strong linkage was formed with a private buyer in Kasese, who procured beans through UNFA and sold to a Kampala-based exporter. In Kibaale the linkage was weaker, due to the inability of BEC to raise finance. This lack of capital was a key constraint, and undermined the efforts that had gone into developing the scheme and building effective outgrower loyalty. Also, competition from other traders emerged for the beans, which affected potential profit. This highlighted the importance of product selection within outgrower schemes and the need to consider diversion factors (whether the product can be used or sold outside the outgrower scheme) and the buyer's (financial) exposure ratio (the likely cost of obtaining the crop against anticipated sales value).

Facilitative Role

Other NGOs and CBOs play a more facilitative role. They assist individuals, groups and communities to market for themselves. This includes both improving access to, and benefits generated from, existing products and existing markets as well as creating new products and new markets (*e.g.* through technology development and processing). There are a variety of ways in which organizations facilitate marketing, including: strengthening the capacity of individuals, groups or communities (through group strengthening and training); developing linkages to traders and other stakeholders in the marketing chain (*e.g.* input suppliers, credit sources and transport agents); and linking farmers to relevant market information. This type of facilitative role is beneficial for a number of reasons: being less interventionist, it is likely to generate more sustainable marketing activities and linkages; it is likely to be achieved at lower cost than if the NGO was more directly responsible for marketing activities; and, therefore, it facilitates reaching a wider audience.

TYPES OF MARKETING INTERVENTION

As indicated in the previous section, there are many different ways in which NGOs or CBOs may intervene to improve access to agricultural markets.

In this section interventions are discussed in eight non-exclusive categories that describe aspects of the intervention strategy:

1. Intended beneficiaries
2. Skills and training
3. Access to agricultural inputs
4. Agro-processing technologies
5. Marketing linkages
6. Credit programmes
7. Marketing information
8. Holistic approaches

Any particular marketing intervention may comprise elements from several categories (*e.g.* inputs and training, or technology, training and finance). The concepts and experiences associated with each category are reviewed, permitting some preliminary conclusions on the more promising strategies.

INTENDED BENEFICIARIES

Individuals, Groups or Communities

Different NGO and CBO marketing initiatives operate with different beneficiary or client structures: some work with whole communities, some with groups and others with households/individuals. The choice of appropriate structure will depend on a number of factors. Research carried out by the Plunkett Foundation and experience of CARE's Development through Conservation project in Uganda suggests that working with village associations or whole communities is more difficult than working with smaller groups. The latter are more focused, more specialized and more likely to have a common goal. Many NGOs and CBOs implement their marketing interventions with groups or associations. Not only are there advantages to the NGOs and CBOs of working with groups, but there may be advantages for the farmers themselves of marketing collectively. The groups can be existing groups (*e.g.* women's groups, savings and credit groups, social groups and so on) or newly formed groups.

The potential advantages of farmers collectively addressing marketing constraints include:

- Economies of scale, through joint purchasing of inputs and joint marketing of products;
- Improved access to finance, where credit organizations favour group loans, or where pooled resources provide the necessary down payment; this can overcome problems of larger investment needed in, for example, processing technologies, storage facilities or transport;
- Collective bargaining power;
- Lower transaction costs (for producers and traders).

Yet there is much evidence that, in general, small enterprises owned and managed by individuals are more successful than group enterprises. Technology adoption in particular is felt to be better amongst individuals than groups, although groups can be an appropriate vehicle for technology transfer if the technologies are subsequently employed by individuals. This is echoed by String fellow *et al.* who argue that group enterprises are more likely to succeed when based on joint marketing rather than joint management/ownership of assets, because the latter requires more complex skills and experience. It has also been found that external organization and management of groups can prevent the development of entrepreneurial skills. There are no definitive rules on which structure is appropriate.

It will depend on the type of intervention and the objectives of the individuals and NGOs/CBOs. On the one hand, NGOs need to remain open to working with groups where appropriate, recognizing the potential to build capacity, reduce transaction costs, and introduce activities with a higher investment threshold. However, an understanding of the reasons behind group formation is important and care should be taken not to over-burden groups formed for social rather than economic reasons. Sometimes individuals, often women and particularly in rural areas, prefer to work together. Furthermore, commonly cited problems attributed to groups per se have been found to be less important when there is strong group leadership and cohesion, and when there is good group organization before any external intervention takes place. String fellow *et al.* point out that a non-interventionist approach – letting producers decide for themselves whether they operate as individuals or groups – allows individuals to develop appropriate structures to build necessary skills and solve their own problems. People will not work well within an imposed structure. As Gibson succinctly states: "Enterprises working in a market environment (whether collectively owned or individually owned) have ultimately to make a surplus in order to survive. The 'bottom line' for NGOs is to support the structures which will work best according to this criterion."

Rural Poor

The majority of NGOs and CBOs deliberately target their activities at poor households and poor communities. Those that focus on agricultural production, processing and marketing are often found in remote rural areas. This focus has implications for the approach adopted by the NGOs/ CBOs and the marketing initiatives they develop. Whereas the trend in marketing approach adopted by NGOs has moved towards a more business-oriented, facilitative approach in recent years, some authors argue that this is less likely to succeed in remote rural areas than in less remote, higher-potential areas. Moreover, within rural communities, some poor individuals and groups are considered too remote and disadvantaged to be able to benefit from marketing interventions. The most vulnerable households and women-headed households in rural areas are particularly disadvantaged; they tend to have more restricted access to information and services and tend to be more risk-averse than other households. Some households may only benefit indirectly from a marketing intervention through any impact it has on the labour market.

Consideration should also be given to the types of intervention most appropriate to a particularly disadvantaged target group, particularly where the transfer of hardware and entrepreneurial skills are envisaged. For example, food processing technologies are often predominantly adopted by the rural elite, due to their more developed entrepreneurial skills and access to finance. This may nonetheless reduce poverty, through improved markets and employment generation, but the strength of these linkages varies, making it difficult to generalize. Certainly more information is needed on the extent to which the poor benefit directly and indirectly from NGO and CBO marketing interventions, and about which approaches and types of intervention are having most impact.

Women In common with other organizations, NGOs and CBOs often find it very difficult to effectively target benefits to women. Yet women are frequently amongst the poorest members of rural communities and face particular constraints in improving their livelihoods.

Moreover, their traditional role in food crop production and (sometimes) domestic marketing, as well as the influence they have over child welfare and nutrition, makes them an obvious target for poverty-focused agricultural marketing interventions. In their review of women's role in post-harvest operations, Gordon *et al.* conclude that:

"Where interventions are intended to benefit the poorest women, attention should be focused on particular issues:

- The needs of female-headed households, which feature disproportionately amongst the poor;
- Crops and processes used in marginal areas;
- Carrying fuel and water, because so many women are affected;
- How poor women earn income – so that new technology really does benefit them;
- Understanding the broader processes which determine how benefits are distributed;
- Household level and informal sector activities, where the poorest people earn their living."

Goodland *et al.* make a number of suggestions on how women's participation in rural finance programmes can be increased, the spirit of which is equally relevant here:

- Out-of-hours opening or mobile services, or locating services in places women frequent, for example, marketplaces;
- Relaxing literacy requirements;
- Flexible collateral requirements, for example, accepting jewellery rather than land;
- Allowing loans of a suitable size (usually small).

SKILLS AND TRAINING

The types of training offered by NGOs in support of marketing initiatives can include group strengthening, general extension, marketing and specialized training.

Group Formation and Strengthening

The advantages of working with groups have already been highlighted. Despite these, group establishment and operation have generally proven more difficult in practice than expected. Some organizations undertake group strengthening and training (*e.g.* in recruitment and management, structure and leadership, financial and business management skills and so on) before marketing activities are initiated, or as part of the marketing intervention. The case of oil palm processing with the NGO TechnoServe is a good example. The danger of not assessing the performance of existing groups and providing any necessary training in group strengthening is that marketing activities implemented with these groups can fail due to general group weaknesses, rather than problems with the marketing activities per se. Stringfellow *et al.* studied farmer co-operative enterprises and their findings highlight the importance of not over-estimating group capacities, and the need for long-term involvement in building group capacities. Often groups fail because they have been formed too quickly and too much is expected of them. They also found that group enterprises are more likely to succeed when based on joint marketing rather than joint management/ ownership of assets which requires more complex skills and experience. It is also important to consider the most appropriate group size and whether there is a culture of working as a group. The Co-operative League of the USA (CLUSA) working with CARE in Mozambique has developed a thorough approach to the development of farmers' associations. They have attempted to make this approach more sustainable by developing the capacity of a local NGO to take-up and continue these activities, and by promoting a structure within the farmers' associations that permits ongoing development. The approach has been criticized for being too costly, but the results to date, albeit over a short duration, are nonetheless very impressive. There is now a need to critically evaluate this approach, to identify the direct and indirect subsidies provided by the NGOs, and assess which components have the most prospects of sustainability.

TechnoServe and Oil Palm Processing in Ghana

TechnoServe is an NGO whose activities focus on the provision of food processing technologies for rural communities. The initial focus of its work was small-scale oil palm processing and extraction in Ghana, but it has subsequently expanded to include processing service centres, inventory credit schemes and production and processing of non-traditional export crops. The purpose of the oil palm scheme was to build on traditional processing (which was laborious and slow) and to exploit the potential for expansion in a strong market, by introducing small-scale mechanized oil mills to community-based groups. The approach adopted by TechnoServe went beyond providing the technology and technical assistance. It was recognized that entrepreneurial skills in the communities were weak, and that group development activities and financial and business training (including linkages to formal credit and extension services) were necessary. This integrated training and support package clearly contributed to the sustainable adoption and management of the oil processing enterprises. However, the Techno Serve approach has also been criticized. The training and support provided to groups are heavily subsidized. Case studies of the oil palm enterprises indicate that the skills are held by only a small number of active group members, and that because of social norms, there is a reluctance to pass these on. A more serious criticism of the approach is that although it has been successful in transferring skills specific to the oil palm enterprise, it has not managed to create more general entrepreneurial skills and characteristics amongst the group members.

Training and Extension

The starting point of a number of CBO/NGO marketing initiatives is production. This is because to market successfully, farmers need to produce and sell what is in demand, at a profit. Often existing markets could be accessible to farmers (either on their own or through linkages with traders), but marketing is constrained by the low volumes or poor quality of farmers' crops. Improved production practices are important for increasing yields of existing crops, new varieties and new crops. Yet government extension services are often lacking or extremely under resourced. NGOs, therefore, often assume a role in providing, or facilitating the provision of, relevant extension information. This may be through strengthening existing extension services or through the establishment of alternative services, such as farmer-extension schemes.

When the CARE Egypt Agricultural Reform Programme began, it focused on improving crop and livestock production. Now, however, the project has become more market oriented and requires farmers to examine the market for their products prior to improving production. Other NGOs and CBOs provide marketing training to groups and individuals. This includes training in production and marketing systems, constraints and opportunities, market demands (products and service) and how to assess whether products can be supplied profitably.

The benefits of providing producers with this type of training are that it develops their capacity to analyse markets for themselves and, therefore, allows them to respond to changing market opportunities and threats. They are not dependent on the NGO/CBO to identify marketing opportunities for them in the longer term. It is also important for NGOs and CBOs not to create parallel services and for different development organizations and governmental bodies to co-ordinate the training and services they provide. In Uganda, for example, a task force was established to prepare guidelines on how to integrate NGO activities into district agricultural programmes. Specialized or vocational training is usually provided to individuals or groups when the marketing intervention relates to a new product or new technology, and requires new skills. Examples include production targeted to a new niche market or particularly quality-conscious export market, or training in the use of an agro-processing technology, such as an oil press. Increasingly organizations which provide skills and training for enterprise development are charging trainees a fee. This helps NGOs and CBOs with limited funding to reach a wider audience. More significantly, it has been found that charging a fee increases the proportion of trainees who actually make effective use of their training. The entrepreneur, making the decision to invest money and time in training, is in effect making a risk assessment.

IMPROVING ACCESS TO AGRICULTURAL INPUTS

Poor access to inputs directly influences the level and quality of production. Even in the poorest parts of Africa, there is still demand for farm implements and good quality seed. Less-poor farmers may make selective use also of fertilizer and pesticides. Input subsidies are a particularly vexed issue. Some argue that they are needed to provide a short-run boost to production and incomes. Yet they are also disruptive and undermining of sustainable commercial development. At a workshop in Uganda, participants highlighted the negative effect of farm input relief programmes in neighbouring countries on the development of commercial input supply networks in Uganda. In Malawi, the starter-pack scheme (distribution of free seed and fertilizer) implemented in 1999 and 2000 has boosted production there, but it has also deprived other low-income farmers in neighbouring countries (notably Mozambique) of a traditional outlet for their surplus production.

Gordon discusses five sets of issues affecting access to inputs: affordability; availability; information; risk and uncertainty; and the overall commercial context. Although credit is often assumed to hold the key to improved access, other ways to improve affordability are also identified: timing input sales to coincide with times when farmers have cash; selling inputs (*e.g.* seed) in small pack sizes suited to small producers; and lowering prices, by making cost reductions in distribution and marketing (*e.g.* through bulk purchases, transport sharing arrangements, and farmers' groups taking on more responsibilities).

Whilst NGOs may play a role in providing credit or promoting some of these other strategies, a role for CBOs (*i.e.* farmers associations) is much more apparent. In countries that have succeeded in significantly improving access to inputs, farmers' associations have acted as a key vehicle for input distribution. Many consider the physical availability of inputs to be a more important constraint to access, with thin and unreliable rural distribution networks in most African countries. A recent development involves innovative approaches to the promotion of input stockist networks by NGOs, illustrating what can be achieved through constructive partnerships between the commercial, private non-profit, farming community and government sectors. These initiatives typically involve training stockists and may involve loans or loan guarantees.

There is also a growing interest in ways to improve and build on traditional informal seed systems. Information constraints are also important – be they in terms of information gaps (basic research on fertilizer response, for instance) or information flows. Although farming is inherently risky, better information reduces uncertainty, enabling farmers to make more informed production decisions. Gordon concludes with some general observations, which also have relevance to the activities of NGOs: "In addition to policies aimed towards the general development of rural economies, a number of more specific policy recommendations are made: avoid actions which undermine the development of sustainable commercial input supply networks; support input markets by setting standards and regulations, and providing information and training; promote synergistic partnerships between commercial, private non-profit, farming community and government sectors; and fill critical research and information gaps."

WTO AND SEED PRODUCTION IN TAMIL NADU

In the present era of WTO it has become absolutely necessary to be prepared to compete in World Trade and any slackness will not only reduce the share in world trade but is also likely to cause heavy damage to the domestic industry through imports. To promote export of agricultural commodities, quantity produced should be surplus after meeting the domestic demands and the quality of the produce should be on par with international standards. The seed is the basic input in Agriculture and it plays a vital role in sustained growth and development of agricultural sectors.

A study report on certified seeds show that 1% increase in production and distribution of certified seeds increases the value of Agricultural production by 1.05% in Tamil Nadu. This shows that agricultural production in the State responds well to certified seeds produced and distributed. At present only about 10% of the total seed requirement in the State is met with certified and 6% with labelled seeds. Thus there exists a wide gap between the requirement and availability.

PROTECTION OF PLANT VARIETIES AND FARMERS' RIGHTS ACT 2001

The objectives of protection of Plant Varieties and Farmers' Rights Act are:

1. To establish an effective system for Protection of Plant Varieties, the rights of the farmers and plant breeders to encourage the development of new varieties of plant.
2. To recognize and protect the right of the farmers in respect of the contribution made at any time in conserving, improving and making available plant genetic resources for the development of new plant varieties.
3. To accelerate agricultural development in the country by protecting the plant breeders' right to stimulate investment for research and development, both in the public and private sector for the development of new plant varieties.
4. To facilitate the growth of the seed industry in the country by ensuring the availability of high quality and planting materials to the farmers.

Patenting of the products may be helpful to improve competitiveness in the International Market. Government of India has preferred to use sui generis system instead of patents because of three major advantages namely flexibility, better protection of farmers' rights and stronger researcher exemption. Accordingly, the Protection of Plant Varieties and Farmers Rights Act was enacted in the Parliament for the registration and better protection of geographical indication of goods.

NINTH PLAN REVIEW

The Ninth Five Year Plan aimed to increase loan facilities to the farmers, to provide transport facilities for transport of produce from village to the regulated markets, to strengthen Agmark laboratories, to create Cold Chain facilities in major marketing centres, decentralized storage facilities for buffer stock of seed production and to improve infrastructure in major rural shandies in each district.

During Ninth Plan 19 Agmark Laboratories were strengthened as against the target of 25 laboratories. Two new Laboratories were constructed as against three laboratories, which were programmed for construction. Ten agricultural production and marketing information centres have been established.

Physical progress made under seed certification, seed testing, and training during the Ninth Plan is given below.

Item / Programme		Units	Ninth Plan	
			Target	Achievement
1	Area registered under Seed Certification	Lakh .Ha	2.12	1.82
2	Quantity of Seeds certified	Lakh MT	1.63	1.97
3	Seed sample tested	Lakh No.	2.30	2.85
4	Seed selling points inspection	Number	112250	130885
5	Seed Samples taken	Number	83000	97697
6	Person trained	Number	10320	125559

TENTH FIVE YEAR PLAN

Approach and Strategy

Agricultural Marketing and Agri-business

- To enhance marketability of agriculture commodities by providing infrastructure facilities, revamping the regulated markets.
- Provision of post harvest handling facilities for value addition and prevention of wastage (like cold storage).
- To provide backward and forward linkages through marketing, agroprocessing and export.
- Better realization for agriculture produce through alternative markets like product-wise Terminal Markets
- Stepping up export of agri/ horti produce-with setting up of AEZ and establishment of Food Laboratories.
- Policy to attract private sector in storage and agro-processing industries. Integrated approach from planting to marketing, which includes choice of crops, grading, packaging, storage and marketing for domestic and international.

Seed Certification

- Increasing the production of certified seed to maintain the quality of produce.
- Improving the storage facilities to preserve the guard samples for a long period.
- Ensuring quality of certified seed distribution among the farming community by strengthening the seed inspection wing of the department.
- Providing computer facilities for improving present reporting system.
- Facilitating timely testing of seeds at reasonable time with modern equipments.
- Establishing market intelligence and demand forecasting cell.
- Creating website exclusively for the Department of Seed Certification.
- Strengthening of Grow Out Test (GOT) Farm at Kannampalayam

TENTH PLAN SCHEMES

Seed Certification Department

Ongoing Schemes

- *Seed Certification Programme (₹ 6 crores):* Seed Certification is a regulatory process designed to secure, maintain and make available the prescribed levels of seed quality namely germination, physical purity, genetic purity and seed health. Though the certified seed production had doubled in a decade, still ten per cent of the total requirement could alone be covered with certified seeds. To increase the area under certified seed production and to increase the quantity of the certified seeds, it is proposed to continue the Seed Certification programme during the Tenth Plan with an outlay of ₹ 6 crores.
- *Seed Inspection (₹ 0.22 crore):* To ensure the quality of the seeds distributed to the farmers, seed selling points are inspected periodically and seed samples are drawn and sent for analysis to the notified Seed Testing Laboratory. Based on the results legal action is being initiated against the defaulters. Quality control measures for seed enhances the agricultural production. This programme will be continued during the Tenth Plan period with an outlay of ₹ 0.22 crore.
- *Seed Testing (₹ 1.75 crores):* Seed testing is being carried out to analyse the quality of the seed lots. Factors like germination; physical purity, moisture, seed health and admixture of other distinguishable varieties are being analysed in the notified seed testing laboratories. There are seven notified Seed testing laboratories functioning in the State with an annual capacity of testing 42,000 seed samples. This

programme will be continued during the Tenth Five Year Plan also with an outlay of ₹1.75 crores.

New Schemes

1. *Creation of Infrastructure facilities (₹4.61 lakhs):* The guard samples pertaining to certification of each seed lot have to be preserved for two years from the date of grant/ extension of the certificate and four years in respect of rejected seed group lots from the date of communication of rejection. It is proposed to equip the office of Assistant Director of Seed Certification with storage facilities at a cost of ₹4.61 lakhs.
2. *Computerisation of offices of the Assistant Director of Seed Inspection (₹8.80 lakhs):* To monitor closely the seed transactions and detain sub standard seeds at once and for the data collection with regard to quantity of seeds distributed and to send the report to the Government, it is proposed to computerise 11 Offices of Assistant Director of Seed Inspection at a total cost of ₹ 8.80 lakhs.
3. *Establishment of Mini Seed Testing Laboratories. (₹1.27 crores):* There are seven notified Seed Testing Laboratories in the State with a capacity to test 42,000 seed samples annually. In view of the increase in the quantum of certified seed production, 62,000 seed samples are being received for testing. To facilitate timely testing of seed samples, it is proposed to establish 13 mini Seed Testing laboratories at a total cost of ₹1.27 crores.
4. *Improve/ Maintenance of the Existing Seed Laboratory Equipment/ Grow out Test Farm (₹13 lakhs):* As a measure to improve the facilities by provision of new equipments in the existing seed testing laboratories and in Grow Out Test Farm at Kannapalayam, a sum of ₹13 lakhs is provided for Tenth Five Year Plan.
5. *Training (₹5 lakhs):* To impart training to the farming community, seed producers and seed dealers, an amount of ₹5 lakhs is provided.
6. *Establishment of Market intelligence and Demand forecasting cell (₹4 lakhs):* To forecast the demand and price of seeds of various crop varieties, establishment of market intelligence and demand forecasting cell in the Directorate of Seed Certification is proposed. An amount of ₹4 lakhs is provided for this programme.
7. *Creation of Web Site exclusively for the Department of Seed Certification (₹0.50 lakh):* It has been proposed to create exclusive website for the Department of Seed Certification to serve the farming community, seed growers, seed producers, seed dealers and distributors at a cost of ₹0.50 lakh.

Agriculture Department

Ongoing Scheme

Establishment of Fertilizer Control Laboratories (₹ 1.08 crores): There are 14 Fertilizer Control Laboratories in the State with an annual capacity to analyse 17,220 numbers of samples to ensure the availability of good quality fertilizers and manure mixtures as per the standards prescribed. An amount of ₹ 1.08 crores is provided for this scheme for the Tenth Five Year Plan.

AGRO-PROCESSING TECHNOLOGY AND MARKETING

Small-scale farming households in remote rural communities generally find that they operate in markets comprising many producers of undifferentiated products, which leads to price competition and low profit margins. Access to processing technology can provide new market opportunities by reducing perishability or adding value in other ways. Processing technologies can range in scale from household-level 'lowtech' processing to fully mechanized factories. Household level processing has two main functions. It can add value and it can preserve the product, thereby increasing the time available for marketing. Other advantages of small-scale agro-processing enterprises are that they can create employment at low levels of investment that make effective use of local resources.

Enterprises owned and managed by individuals or households are often more successful than group enterprises, so technology development organizations need to be aware of this need for small-scale technologies. However, many NGOs and CBOs work with farmer groups either because it is the structure that the farmers prefer (there may be a culture of group activity) or because the NGO/CBO prefers this approach (due to financial and coverage reasons). Yet group approaches to the adoption of agroprocessing technology are often weak and entrepreneurial skills are less evident than when working with active individuals. In these circumstances, careful group selection is required, as well as consideration of ways in which the need for entrepreneurial skills can be reduced by introducing a third party, such as a private company offering marketing services to small-scale processors.

It is necessary to ensure that there is market demand for the technologies and/or their end-products and that the technologies are appropriate (*e.g.* taking into account gender issues), rather than developing technologies for their own sake. It is also important that the wider enterprise environment is considered and that institutional arrangements enable smallholder access to markets. Also, as will be examined further, many marketing constraints faced by farmers are interrelated. In terms of technology development, other marketing factors such as access to credit, adequate managerial and technical skills, and market information all influence technology choice decisions.

The Ram Press – Oilseed Processing Technology Introduced by ATI

In the 1980s, ATI (now Enterprise Works Worldwide) started developing a manual ram press (also called the Bielenberg press after the engineer who developed the prototype) to produce edible oil in remote parts of Africa without electricity and with poor access to markets. The first presses were intended for soft-shelled sunflower seed in Tanzania, but the model has since been adapted for use with other oilseeds, including coconut and sesame. The first presses were arduous to use but later models improved on this and could be operated easily by one person. It was hoped that they would improve nutrition by improving access to energy-rich food, whilst also increasing farmer incomes and creating enterprise and employment in rural areas. Indeed, local entrepreneurs involved in small-scale oilseed processing can earn two to three times more gross income, compared with selling the seed to processing factories. In Tanzania, costs could be recovered in just one 3-month season. ATI worked up the promotion methodology over many years in many African countries.

The ram press is quite simple and is supplied with a filtration device and tools for maintenance. Training, information and support on proper use and socio-economic and nutritional benefits of an oilseed processing operation are important elements of the extension programme adopted by ATI and its partners. In Tanzania, ATI sold the presses on credit for many years. Elsewhere, whilst varying in their systems and repayment rigour, credit has been an important factor affecting uptake. Depending on the model and manufacturer, the presses cost between US$ 150 and US$ 300. ATI has been innovative in promoting the commercialization of small-scale oil processing in Africa with its Regional Oils Programme. Its focus shifted from providing technical assistance to NGOs involved in small-scale oilseed processing to a private sector approach that emphasizes commercialization and mass manufacturing of ram presses. As a result, private enterprises have been developed, given assistance with market surveys and business plans, and are becoming responsible for the manufacture and wholesale of ram presses. ATI also stress the importance of the participatory processes used that involve economic actors in the community – owners and workers, press manufacturers, sales agents and oil-consuming families.

MARKETING LINKAGES

Marketing problems identified by producers are often attributed to the commercial sector and the capacity to access it: lack of buyers, unreasonably low farm-gate prices, inflexible requirements, and so on. Links between NGOs/CBOs, the governmental sector and commercial agents are usually weak. Yet working together with the private sector is an important way for farmers to access relevant market information, technologies and new market opportunities.

It is also a way for NGOs and CBOs to reach a wider audience with limited funds. As Kleih has stated, although NGOs are making a very positive contribution to rural development, they generally only reach around 1–2% of farming households in any one country. Private sector agents are sometimes willing to collaborate with NGOs and farmer groups to share the costs of providing training and information, if they see it as an investment through which they can increase their own revenue. In these circumstances, both the producers and traders can benefit. Developing and strengthening relationships between farmers and traders can reduce transaction costs, transport costs and risk on both sides. It was noted above that approaches that use and strengthen existing private sector production and marketing channels, rather than seeking to override them or invent new ones, have been particularly effective in addressing marketing constraints and in sustaining linkages beyond the life of the NGO marketing project. The Smallholder Agribusiness Development Promgramme (SADP) in Malawi, established by the American Co-operative Development Initiative and Volunteers in Co-operative Assistance (ACDI/VOCA), links farmers to existing private-sector markets, rather than establishing new market channels. This has contributed to its success in working with large numbers of farmers and achieving sustainability. CLUSA and CARE have also focused on developing linkages between farmers and traders in Mozambique.

CREDIT PROGRAMMES

Credit merits particular attention in the context of agricultural marketing and processing. Farmers groups and NGOs often recognize a lack of credit as a critical constraint to the development of new initiatives and many seek to remedy this through credit interventions. The aspects discussed here are drawn from a recent synopsis by Gordon.

Rural Finance

Rural finance comprises credit, savings and insurance (or insurance substitutes) in rural areas, whether provided through formal or informal mechanisms. The word 'credit' tends to be associated with enterprise development, whereas rural finance also includes savings and insurance mechanisms used by the poor to protect and stabilize their families and livelihoods (not just their businesses). Rural finance comprises informal and formal sectors. Examples of formal sources of credit include: banks, projects and contract farmer schemes. Reference is often made to micro-credit. Micro underlines the small loan size normally associated with the borrowing requirements of poor rural populations, and micro-credit schemes use specially developed pro-poor lending methodologies. Rural populations, however, are much more dependent on informal sources of finance (including loans from family or friends, moneylenders, and rotating or accumulating savings and credit associations). There is an enormous literature on rural finance and micro-credit, much of which relates to small-scale enterprise. (Many micro-credit schemes specifically exclude agricultural production activities, even in rural areas, because they are considered high risk.) The discussion here focuses on particular aspects of rural finance that are relevant to NGO and CBO agricultural marketing initiatives. First, typical features of micro-credit schemes, which tend to be run by NGOs and often work with community groups, are reviewed. Attention is then focused on two types of credit scheme with particular relevance to agricultural marketing interventions: inventory credit – a relatively new departure in smallholder agriculture, which is attracting increasing attention from NGOs and private banks; and outgrower schemes, which tend to be supported by the private sector but often work through producer groups. Some NGOs support other 'hybrid' rural finance initiatives that may affect agricultural marketing. These include stamp-based savings groups and rotating savings and credit associations (ROSCAs), where pooled savings may facilitate access to additional loans, and permit crop purchase/assembly/marketing activities or improved access to agricultural inputs.

Micro-credit and the Rural Poor: The Issues

Rural credit would not be the focus of so much development effort were it not for widespread market failure (*i.e.* failure to provide the services people want) in rural financial services in developing countries.

Reasons for market failure include:

- The lender does not know the default risk of each potential borrower and to collcct this information is costly;
- It is costly to ensure that the potential borrowers take those actions which make loan repayment more likely;
- It is difficult and costly to enforce repayment;
- The cost of providing services to the rural poor is high because they are located in remote areas, want to borrow small amounts, and illiteracy, lack of experience of banks, and lack of collateral necessitate the development of tailored approaches.

What are the implications of this for agricultural activities? Firstly, all these types of market failure apply to agricultural lending in developing countries. Lack of information on the risk of default is particularly germane to agricultural enterprise. Farmers do not keep records, so it is difficult for them to produce the information that might convince a bank of their creditworthiness. Rural market transactions are largely informal, so it is difficult for the bank to collect independent information on prices. Farming is clearly a risky business because of weather, pests and market fluctuations and it is difficult for a bank to assess the degree of risk associated with particular activities. The rural poor do not have a track record, or referees who will vouch for their competence and reliability. Making sure that farmers keep to their business plan, using loans as intended, and carrying out tasks to schedule, is also costly – although this might make loan repayment more likely. Enforcing repayment is also difficult. This requires monitors who know when crops are sold, or agreements with merchants to pay the farmer net of what she/he owes the bank, or effective penalties such as seizure of assets or prosecution. Farmers rarely have collateral acceptable to banks.

They may not have clear title to the land they farm, or even if they do, rural land markets may not function well enough for land to be considered a 'bankable' asset. Poor farmers, moreover, rarely have other bankable assets. They might own a bicycle, and have a store half full of grain, but were a bank to seize such assets the cost of doing so would probably exceed their sale value. The poorer the farmer, the fewer are his/her chances of borrowing from the formal sector. Women, particularly poor women, face even more problems in obtaining credit. Land title, where it exists, is usually held by men. Women often have little control over other factors of production, particularly for the 'bankable' cash-cropping activities. In some countries women may only borrow in the names of their husbands, if at all, and literacy rates for poor women are almost always lower than those for men. The irony is that numerous studies show that women tend to repay loans more reliably than men. Numerous projects, government schemes and NGOs engage in loan programmes targeted at the poor. Some of these work well, but many are unsustainable because of high and subsidized costs, and high rates of default. Moreover, many miss their target, with the benefits captured by the less poor. The poor depend overwhelmingly on the informal sector. Micro-finance is the response to market failure in 'conventional' banking services for the rural poor. It responds to the needs of low-income households.

Sound schemes tend to be characterized by:

- Small, short-term loans, and savings mechanisms;
- Simplified loan appraisal procedures;
- Innovative approaches to collateral;
- Rapid approval/disbursement of repeat loans after repayment;
- High transaction costs;
- High repayment rates;
- Savings and loan services provided at a location and time convenient to the poor.

Thus, micro-credit schemes are often associated with: group-lending (where peer pressure effectively substitutes for collateral, and other group members may take action to prevent one member defaulting, for instance, by providing labour to assure timely harvest); extension inputs arranged by the micro-finance institute (MFI); and mobile banking arrangements. Cashflow analysis may concentrate on overall ability to repay the loan rather than a particular investment project. In some respects, MFIs try to imitate the strengths of the informal sector (using local information to ensure repayment, for instance) and some MFIs are experimenting with ways to link their operations with some of the informal sector financial agents.

Agricultural marketing loans are not common in MFI loan portfolios, but some micro-credit schemes do provide marketing finance or input loans. MFIs can include government and commercial banks, NGOs and savings and loan co-operatives. Some of the largest MFIs are Asian (for instance, Grameen Bank in Bangladesh has 2 million borrowers and savers, Bank Rakyat Indonesia/Unit Desa system has 2 million borrowers and 12 million savers, and the Bank for Agriculture and Agricultural Co-operatives (BAAC) in Thailand has 2.5 million clients). Coverage is less in Africa, but still growing. K-Rep in Kenya has 15000 clients and the Fe´de´ ration des Caisses d´ e´pargne et de Cre´ dit Agricole Mutuel (FECECAM) credit movement in Benin has 200 000 clients. Whilst these figures are impressive, they indicate a major difficulty: there are too few services to go round – and still fewer likely to become sustainable. Many schemes find it difficult to graduate from subsidized operations to full cost-recovery. Uncosted but critical inputs by NGO staff are commonplace. Interest rates are often set below market rates (even at negative real rates, where high inflation prevails). Too little attention may be focused on repayment and follow-up; and farmer experience of subsidies, free inputs and loan amnesties may all contribute to low repayment rates.

Contract Farmer Schemes

Contract farmer or outgrower schemes focus on very specific needs. They usually operate in situations where a processor or trader faces a supply constraint and, therefore, wishes to promote production of crop x, and has access to sufficient resources (own or loaned) to do so. Inputs are usually provided in-kind (to reduce diversion to other activities), accompanied by extension, and the cost of these services is recouped out of the price paid to the farmer when the crop is harvested. The degree of supervision and type of inputs provided varies greatly. High value horticultural products produced for export may be closely supervised, for instance, whilst supervision of a cotton crop (which still has to undergo considerable processing before it reaches the consumer) may be minimal. Viable schemes must include mechanisms to minimize on-farm consumption or side-selling (farmers selling their crop to an alternative buyer and, therefore, avoiding loan repayment), such as:

- Sharing of information and co-operation amongst crop buyers;
- Effective penalties against farmers who default (exclusion from future schemes, seizure of assets, prosecution);
- Peer pressure achieved through group lending.

Some of the longer-established cotton schemes in West Africa are wide ranging in scope. Considerable long-term effort has gone into capacity building of farmers' groups. This enables farmers to take on a greater share of tasks in input supply and crop assembly, which both reduces cotton company costs and increases cash revenues to farmers. Inputs and extension are available for other crops in the farming system (not just for cotton) effectively reinforcing and expanding the benefits accruing from cotton production.

These types of schemes are common for smallholder annuals grown for export such as tobacco, cotton and horticultural production, where the on-farm investment costs are relatively low, and the pay-back period short. (Thus, greenhouse production of flowers for export in Zimbabwe is not a smallholder activity.) In Africa, the cotton schemes are undoubtedly the largest. About 300000 Ugandan farmers benefit from an input scheme organized by the cotton ginners. In Zimbabwe, around 60000 communal farmers take input loans from cotton companies (and many more participate in other schemes intended to increase cotton output). In Mali, around 100000 rural households participate in the cotton input schemes. The schemes in Mali and Zimbabwe both depend on a network of strong farmers groups, and the Ugandan scheme is likely to develop in that direction as it matures, seeking less costly ways to reach its target group.

Crop Market Characteristics

If there is limited on-farm consumption and local marketing, it may be possible to collect loan repayments at the point of sale. However, unless there is a crop purchase monopoly, or agreement to share information between different buyers, the farmer may be able to avoid repayment by 'side-selling' (selling to another buyer).

Input qualities

Inputs are often provided in-kind to reduce 'diversion' of the input away from the targeted crop. Diversion is lessened if there is a limited alternative use or market for the inputs, or if (unusually) returns to use of the input are greatest for the crop in question.

Commercial/Credit Context

Prospects for viable operation of the scheme are greater if farmers treat farming as a business and are integrated into markets, and if there are supportive legal/political/ contract enforcement institutions. A recent history of loan amnesties, default without penalty, and subsidized inputs may undermine the operation of viable schemes.

Modus operandi of scheme – best practice:

1. Group schemes for peer pressure.
2. Group or individual schemes backed up by monitoring/good information, support staff, and ability to act.
3. Incentives for repayment and penalties for non-repayment.
4. Appropriate incentives for field monitors/co-ordinators.
5. Training provided to farmers extension and business management.

6. Developing relationship/trust/loyalty through field presence/contact.
7. Accessibility of scheme – minimize red tape and transaction costs; organize so that the location and timing of contact is convenient to farmers.
8. Effective and timely monitoring of input use and crop marketing.

Inventory Credit

Another mechanism for financing agricultural trade exists in inventory credit or warehouse receipt systems. This involves a tripartite agreement between a bank, a borrower (usually a trader), and a warehouse operator. Essentially, a trader is able to use an existing stock of, say, grain as collateral for a bank loan providing certain conditions are met. The grain must meet certain verifiable specifications and be stored in a warehouse operated by a third party. The bank will usually lend up to a certain percentage (perhaps 80%) of the value of the grain at that time (at harvest time when prices are low). The trader can then use the loan to acquire additional stocks of grain, and is usually required to repay the loan during the 'lean' season before grain prices start to fall once more with the next season's harvest. The trader must settle all the warehousing costs before she/he removes the grain, and is generally required to repay the loan at this stage also. In the event that she/he is unable to repay the loan, the bank can seize the grain and sell it.

There is a lot of interest in inventory credit at the present time because of the scope it offers to provide traders with capital to fuel agricultural marketing. However, its use is limited by a number of factors: it can only be used for crops that are relatively non-perishable, with reasonably predictable pricing scenarios; by definition it is targeted to larger, more sophisticated traders, able to purchase initial stocks and fairly confident in their negotiations with banks and warehouse operators; and it is critically dependent on the necessary institutional infrastructure (bank willingness to lend against inventories, warehouse systems which can operate to the standards and within the necessary legal framework, and supportive and enforceable legal institutions). NGO involvement in these schemes can take a number of forms. It may include a role as an MFI providing the credit, or it may include working with farmers' groups who effectively become the 'trader' in the tripartite agreement, buying crops from other farmers (or their membership), for storage and sale later in the season when prices are higher.

Who Benefits from Rural Credit Schemes?

Particularly those of the poor, tend to be barred from formal sources of credit. Lack of capital is a widely recognized constraint to marketing and enterprise development. Both producers and traders need capital. It is needed to purchase inputs, to allow farmers greater flexibility in the timing of sales and to allow traders to engage in intra-seasonal storage. In recent years there has been significant growth in the provision of micro-credit services to help businesses develop. Although there has been reasonable success in terms of widespread uptake (particularly amongst women) and high loan-recovery rates, the impact has been relatively disappointing.

Table. Who and what is Excluded or Less Excluded from Rural Credit

Tending to be excluded ←	→ Tending to be less excluded
Formal sources of credit	
Women	Men
The poor	Less poor
Illiterate	Literate
Agricultural activities	Non-farm enterprise
Landless	Those with land
Those with no bankable assets	Those with other collateral
Food crops	Cash crops – especially export crops
Unforeseen urgent needs	Planned investments
Informal sources of credit	
Long-term	Short-term
Large loan size	Small loan size
Seasonal inputs and investment	Crisis loans

One of the reasons for this is that much of the credit has been used for basic trading and simple processing activities, which provide limited scope for increased productivity and added value. There has also been relatively little use of credit for technology development. This has led to an expansion of similar enterprises, producing a limited number of goods and services for an already saturated market. A second reason is that small producers and small enterprises cannot afford to take the risk of developing, testing and promoting a new product. They tend to be risk-averse because of limited financial resources and because they lack access to the necessary information to help them profitably change their businesses.

MARKETING INFORMATION

Market information generally refers to market price information, and in some cases includes information on quantities. Marketing information is a wider concept, including information on marketing channels, buyers, quality standards and so on. Post-independence, many market information systems (MIS) in developing countries were government-run but most are now either defunct or confronting major problems. Yet accurate, appropriate and timely marketing information (on prices and on marketing issues more broadly) is very important for producers and traders. Market information informs production, processing, storage and marketing decisions, facilitates the spatial and temporal distribution of products and can strengthen the bargaining power of producers. Yet in practice, MIS have repeatedly proven to be unsustainable or have failed to provide timely and useful services. This has led some authors to argue that users should pay for market information. However, others argue that in the context of small-scale, resource-poor farmers in remote communities, payment for marketing information is unrealistic and information should continue to be considered as a public good. This is particularly true for information provided using mass-media, which makes cost-recovery very difficult. Also, farmers in remote areas with poor coverage by traders often point out that knowing market prices does little to improve their bargaining position if there is little competition amongst traders. The MIS in Mali, which was reorganized in 1998, illustrates some more positive features. The objectives of this reorganization were to "create a decentralized MIS which is efficient, viable and sustainable". A participatory needs assessment was carried out initially to identify the marketing information needs of different stakeholders.

Private sector stakeholder involvement and their needs are recognized as one of the key features of the MIS. One of the objectives of this MIS is to recover some of the costs through the sale of information to commercial stakeholders. However, for remote areas, the information is viewed as a public good, and it is not anticipated that the system will be able to work without donor support. NGOs may have a role to play in collecting and/or disseminating market information where local government or private sector capacity is weak. It can involve assistance to local government in establishing MIS, training information officers and/or provision of transport and equipment.

However, it is important that the institutional side of any established MIS is sustainable; the creation of new and unsustainable units, which do not form part of the local institutional set-up, should be avoided. Also, it is important that MIS are co-ordinated. The different marketing information requirements of producers and other stakeholders in the marketing chain may require several different sources of information (*e.g.* NGOs and CBOs, government extension and research services, private companies and so on). However, co-ordination is required to avoid duplication and to use scare resources efficiently.

HOLISTIC APPROACHES

Numerous authors have indicated that marketing constraints cannot be tackled individually. The experience of many NGOs and CBOs has shown that isolated marketing interventions, for example, the provision of new processing technology or market information, are unlikely to succeed. A holistic approach is needed to marketing and enterprise development, to look at the whole range of marketing constraints so as to improve the terms on which farmers participate in the market. Indeed, this is the strength of the private sector; it adopts a holistic approach when assessing the feasibility of enterprises and balances market demands, technology, infrastructure, training and other requirements. Many of the marketing interventions already described in this review have been successful through adopting a holistic approach. For example, the Appropriate

Technology oil palm press initiative involved much more than just design and provision of the technology, likewise the CLUSA approach to farmers associations. Training was provided to recipient groups in production, marketing, technical aspects of the oil press including repairs and agribusiness management, and linkages with manufacturers and traders were facilitated. The Smallholder Agribusiness Development Programme (SADP) in Malawi is another good example of a holistic marketing programme.

ACDI/VOCA – Addressing a Range of Marketing Constraints in Malawi

The Smallholder Agribusiness Development Programme (SADP) was established by the American Co-operative Development Initiative and Volunteers in Co-operative Assistance (ACDI/VOCA) and its primary aim is to improve market access for farmers. The programme was initiated to work with tobacco farmers (the largest and most profitable cash-crop sector in the country) at a time when the sector was being deregulated in the early 1990s. The programme works with existing clubs or associations of tobacco farmers, who are responsible for bringing farmers together to jointly solve their problems. The programme deliberately selects stronger functioning farmer clubs to work with and they are expected to operate as commercial businesses. The services provided by the programme are broad and address a range of marketing constraints.

They include:

- Capacity building of groups and associations (through management and financial training);
- Assistance in identifying problems;
- Assistance in identifying and developing solutions by:
  - Providing training and market information
  - Facilitating linkages, identifying private sector service providers
  - Negotiating transportation and other service agreements
  - Developing improved delivery systems.

The programme strives to be sustainable through a number of measures:

- Farmers contribute financially through club association membership fees;
- Clubs and associations do not provide services directly (this would need more capital and management training) and are linked to commercial credit sources, rather than receiving grants;
- Where possible, services are contracted out to the private sector;
- Establishing a national organization (NASFAM) to provide these services beyond the life of the programme.

A number of the marketing services have proven profitable and sustainable (11 of the 12 farmer associations were financially sustainable in 1998). Those that have not been include the collection and dissemination of marketing information, auditing and monitoring performance of farmer associations and helping farmers to diversify their crop base.

CONCLUSION

The argument in favour of agricultural and food enterprises, in developing countries, becoming more customer oriented is a compelling one. Average incomes in developing countries are low and so the marketing systems which deliver agricultural and food products have to be efficient if they are to deliver food and other products at affordable prices. Moreover, when a country does experience economic growth this is normally accompanied by an acceleration in the rate of urbanisation. The end result is that greater demands are placed upon farmers.

Marketing systems have to be capable of signalling the needs of both consumers and industrial users of agricultural outputs to the farmers. The marketing system must also motivate and reward all of the parties whose participation is essential to the delivery of commodities and products in the quantities and at the qualities demanded. Yet another development which has increased interest in marketing practices of late is the move towards market liberalisation as part of economic structural adjustment in many developing countries.

The marketing concept suggests that an organisation is best able to achieve its long term objectives by orientating all of its operations towards the task of consistently delivering satisfaction to the customer. In order to do so, the organisation must begin by getting to know what it is that will satisfy the customer. The marketing system as a whole has to be customer orientated. A marketing system comprises the functions of marketing (buying and selling, storage, transport and processing, and, standardisation of weights and measures, financing, risk bearing and market intelligence), and the organisations that perform them. Marketing systems have at least four sub-systems, these being production, distribution, consumption and regulation. These sub-systems often have conflicting interests that have to be resolved if the system as a whole is to be efficient and effective.

The food industry is a major user of agricultural products and commodities. As disposable incomes increase in developing countries, the food industry will have to meet new and different needs from its more affluent consumers. The food industry will, in turn, require agriculture support its efforts to meet the new challenges and opportunities. In particular, the food industry will demand that agriculture produces a wider range of qualities in its products and commodities with a greater proportion of total supply in the top grades; downward pressure will be exerted on agricultural production costs; agriculture will be required to supply throughout the year rather than seasonally; reliability in the quantity, quality and timing of supplies will become the major determinant in supplier selection; innovative producers who can provide differentiated products and products that make food processing easier or cheaper are more likely to survive than those who persist in producing traditional products using traditional farming methods; and issues related to the health aspects of food consumption will become increasingly important.

Bibliography

A. Poshadri and Aparna Kuna : *A Handbook of Food Techno's*, New India Publishing Agency, Delhi, 2013.

A.D. Dholakia: *Delicious Seafood Recipes*, Daya Publications, Delhi, 2010.

Andrew L. Winton and Kate Barber Winton : *A Handbook of Structure and Composition of Foods*, Agrobios Publication, Delhi, 2013.

Ashok K Chauhan and Ajit Varma:*Microbes : Health and Environment*, I K International, Delhi, 2007.

Aysha Aziz Faridi:*Dairy Microbiology*, Random Publications, Delhi, 2012.

Cristobal Noe Aguilar; Juliana Morales Castro; Efren Delgado; Diana Jasso Cantu and Ashok Pandey: *Food Science and Food*

Dalip Kumar and Asmi Raza: *Agriculture and Food Security*: *Contemporary Issues*, Deep and Deep Publications, Delhi, 2011.

Doreen Virtue : *Constant Craving : What Your Food Cravings Mean And How To Overcome Them* , Hay House India, Delhi, 2011.

Harmeet Singh: *Dairy Farming*, APH Publication, Delhi, 2011.

Hema Thapar: *Food Science and Health*, Pacific Publications, Delhi, 2011.

Hema Thapar: *Nutrition and Food Science*, Pacific Books International, Delhi, 2011.

M Lakshmi Narasaiah: *Economic Growth and Food Security*, Discovery Publishing House, Delhi, 2008.

Madhulika Bhatnagar.: *Encyclopaedia of Catering Technology, Food Service and Hospitality Management, Vol. I to III*, Delhi, Anmol Publication, 2008.

Manju Malhi: *Easy Indian Cookbook: The Step-By-Step Guide to Deliciously Easy Indian Food at Home*, Viva Books, Delhi, 2008.

Manoj Negi: *Career in Dairy Farming*, Abhishek Publication, Delhi, 2010.

P R Gupta:*Dairy India 2007: Serving the Dairy Industry Since 1983*, Dairy India Yearbook, 2007.

P T Rajan : *A Field Guide to Marine Food Fishes of Andaman and Nicobar Islands*, Zoological Survey of India, Delhi, 2003.

P.C. Trivedi:*Microbes : Applications and Effects*, Aavishkar Publication, Delhi, 2009.

Piyush Sharma and Prem Ram : *A Textbook of Food and Beverage Management*, Dominant Publication, Delhi, 2013.

R S Paroda: *Sustaining Our Food Security*, Konark Publication, Delhi, 2003.

R. P. Saxena.: *Hotel Management : Food And Food Services*, Delhi, Centrum Press, 2010.

S.K. Singh: *Fundamental of Hotel Management and Operations*, Delhi, Centrum Press, 2010.

S.K. Singh: *Strategic Management in Hotel and Restaurant Industry*, Delhi, Centrum Press, 2010.

Tapeshwar Singh: *Resource Conservation and Food Security : An Indian Experience (2 Vols-Set)*, Concept Publication, New Delhi, 2004.

Tharakan : *A Young Hotelier's Guide to Food and Beverage*, Tata McGraw-Hill Publication, Delhi, 2005.

Udai Veer: *Elements of Food Science*, Anmol Publications, Delhi, 2007.

Urvashi Nandal : *A Handbook of Foods and Nutritional Biochemistry*, Agrobios Publication, Delhi, 2013.

Vishwambhar Prasad Sati: *Natural Resources Conservation and Food Security*, Bishen Singh Mahendra Pal Singh, Dehra Dun, 2012.

W. Merback and A. Veha : *Crop Science and Technology for Food Security Bioenergy and Sustainability*, L. Bona, J. Pauk, ,

W.L. Slatter, T. Krishtofferson and D.A. Seiberling: *Manual of Dairy Industry : Training and Development of Personnel for the Dairy Industry*, Asiatic Publication, Delhi, 2012.

Index